The Boomerang Economy

The Boomerang Economy

Why British offshored manufacturers are returning home and how to maximise this trend

David Merlin-Jones

Civitas: Institute for the Study of Civil Society
London
Registered Charity No. 1085494

First Published July 2012

55 Tufton Street
London SW1P 3QL

email: info@civitas.org.uk

ISBN 978-1-906837-41-9

Independence: Civitas: Institute for the Study of Civil Society is a registered educational charity (No. 1085494) and a company limited by guarantee (No. 04023541). Civitas is financed from a variety of private sources to avoid over-reliance on any single or small group of donors.

All publications are independently refereed. All the Institute's publications seek to further its objective of promoting the advancement of learning. The views expressed are those of the authors, not of the Institute.

Typeset by
Civitas

Printed in Great Britain by
Disc To Print
London

Globalization is a policy, not an act of God.

Jimmy Carter

Contents

Author

David Merlin-Jones, FRSA is a graduate of Exeter College, Oxford University where he studied History. He joined Civitas in 2010 as a Research Fellow, focusing on economic issues and, in particular, British manufacturing and energy. He is now Director of the Wealth of Nations Project. He has previously authored *Chain Reactions: How the chemical industry can shrink our carbon footprint* (2011), *CO2.1: Beyond the EU's Emissions Trading System* (2012) and *Extending Lending: The case for a state-backed investment bank* (2012).

He has published other Civitas reports including *The closure of the Lynemouth aluminium smelter: an analysis* (2012), *Rock Solid? An investigation into the British cement industry* (2010), and co-authored *A Strategy for Economic Growth* (2011). His work has been discussed on the BBC, including the World Service, and in *City A.M., The Guardian, The Financial Times, The Sunday Times* and *Telegraph* amongst others.

He is also a Fellow of the Royal Society for the Encouragement of Arts, Manufactures and Commerce.

Acknowledgements
The author would like to thank Georgia Plank, researcher at Civitas, for her assistance in completing this publication. In addition, he would like to thank the referees of the report for giving up their time and offering their expertise: Steve Hughes, Senior Economic Adviser at the British Chamber of Commerce, Marcus Gibson, editor at Gibson Index and David Bailey, Professor of International Business Strategy and Economics at the University of Coventry. Any remaining mistakes are, of course, his own.

Executive summary

- Onshoring in a sustained and predictable manner is a new phenomenon that will increase in the coming years. The Government has a crucial role to play in encouraging this.
- Manufacturers can offshore either by moving their own production abroad, or by outsourcing work to overseas firms. They onshore by relocating production back to Britain or by outsourcing to a British company.
- Offshoring was driven by the desire to cut down production costs to increase profits or simply to survive, mostly through the cheap labour on offer in emerging economies. A 2008 survey found that over a fifth of manufacturers had offshored production to replace UK capacity and another fifth were planning to do so in the future. Over half were going to create offshore plants to increase their capacity and two-thirds also claimed to have outsourced some component manufacturing abroad, with a further fifth planning likewise.
- Some companies started overseas operations as a means to access foreign markets. This works in tandem with British production, and is less viable for onshoring.
- Onshoring will benefit the UK via an increase in employment and a reduction in our trade deficit via increased exports and import substitution. Intangible benefits also include greater self-

reliance, an entrenchment of existing supply chains and improved informal innovation.

- This report examines onshoring of manufacturing from China specifically, as this is where the bulk of British offshorers ended up. Companies could relocate from China to other lower-cost countries such as Vietnam and the Philippines, but many find that the poor infrastructure and skills available in these countries prevent this, making onshoring back to the UK the likely alternative.

Why Chinese production is less attractive

- The labour cost savings, a principal reason for many companies to offshore, are diminishing. The rate of yearly growth in Chinese wages from 2002-08 was 33 per cent per annum. Taking the recession into account, Chinese wages grew by 19 per cent per annum from 2005-10. In comparison, the UK's wages actually shrank by 0.2 per cent over this period.
- As labour constitutes a decreasing proportion of production costs, while Chinese wages will still be a sixth to an eighth of British ones in 2015, actual costs saved through cheap labour are only to be 17-18 per cent of total production costs. This is not enough to keep manufacturers offshoring, especially given rising costs elsewhere.
- It is not as easy to recruit a sizable workforce in a short period of time as it was a few years ago. The general rise in wages has meant fewer migrants to the East Coast, as well-paid jobs can now be found inland. This drives up wage costs further, and in

some major offshore areas, there are actual worker shortages. Guangzhou estimated it faced a labour shortage of 150,000 employees in 2011.

- While labour is cheaper inland, the infrastructure of central China is poor, and it can take almost as long for goods to arrive at Chinese ports as it would to ship them from the coast to Britain. This is therefore not an option for some companies.
- Maintenance of quality is difficult, especially for outsourcers. Given the frequent lack of onsite presence, problems are often discovered only when the product is delivered, which can render a shipment useless while other faulty goods are *en route*.
- Reliability is often an issue, in part due to the long distance of shipping, but Chinese firms are often unable to deliver their goods exactly when needed. This is especially an issue for firms that rely on just in time or lean manufacturing methods.
- In January 2000, the nominal spot price for European Brent crude oil was $25.51 a barrel and the cost of shipping tripled by 2008 when an oil spike occurred. This saw crude prices at a high of $132.72 a barrel and prices look set to reach similar levels again, leading to similarly expensive shipping costs: in February 2012, prices hovered around $120 a barrel.
- The strengthening of the yuan and simultaneous weakening of the pound mean that depending on how British firms pay for Chinese goods, they are

paying higher prices for the same goods. This trend is only likely to continue.

The example of the United States

- The US has experienced and encouraged onshoring for longer than the UK, and provides an example of how Britain can best cultivate the trend.
- Like the UK, the US offshored a sizeable proportion of industry, with American manufacturing multinationals employing two-fifths of their workforce outside the States. Since the recession, the US has seen its manufacturing employment grow faster than any other developed economy and in 2010 it gained 300,000 industrial jobs.
- American onshoring has been driven by depressed wages in manufacturing areas, flexible unions, and arguably most importantly, very explicit support from the government. Obama made onshoring a central part of his 2012 State of the Union address and this has been reinforced at a federal and state level through funds and active intervention for onshorers.
- The trend is being somewhat held back by concerns that the general manufacturing workforce is not equipped to deal with the advanced machinery onshorers prefer to use. A modern, IT-enabled skillset is required by all employees.

Why UK production is increasingly attractive

- Essentially, the machines used in production will cost the same be it in the UK or elsewhere, and the main variable running cost is then labour. Depending on the level of automation, this might be of little consequence and if a company can reduce production time and increase efficiency, this matters even less and could even nullify the Chinese advantage.
- As a result of these advanced production processes, British production is particularly viable for companies producing high-value goods. This value can come from the quality of the good itself or from other aspects such as customisation for the customer and the addition of services. The 'made in Britain' label is the antithesis of the 'made in China' label. It has connotations of high quality, reliability and durability. All these factors mean international customers are willing to pay a premium for UK goods, so manufacturers can compete in markets China and offshorers cannot.
- The old manufacturing heartlands offer cheaper labour at around 10-15 per cent less than the UK average wage. In the Chinese coastal cities, the most desirable locations for offshorers, wages are at a 10-15 per cent premium.
- Using a domestic supply chain ensures problems can be dealt with swiftly and even pre-empted. Close geographical proximity also enables greater innovative capacity and harmony between suppliers and customers, to the benefit of all in the supply chain.

How to increase the volume of UK onshoring

- It is clear that China is no longer the default option when a British company is looking for a location from which to produce goods, but it would be wrong to think this means Britain is now the default choice instead: it is not. Businesses still need to be motivated into retaining British production or repatriating it.
- Relocating back to the UK will be a heavy investment for most companies, so they will want a firm commitment from the Government that manufacturing is valued and viable in Britain for the long-term.
- Improve skills: the example of America tells us that a shortage of the right skills is a bottleneck for onshoring. Too many potential workers do not have the ability to work with modern, computerised equipment without significant training. There is also too much emphasis on school leavers choosing academic qualifications over higher level vocational ones. Companies also need employees with actual industrial experience, which is increasingly hard to come by in graduates.
- Active assistance: America has managed to secure the return of many onshorers through the liberal use of providing financial incentives. Electrolux recently received a total of $188 million through various funds and benefits to set up a new $190 million plant in Tennessee. The UK needs to restart the policy of active intervention it enacted so well under Thatcher. This also sends a

confident message of support to manufacturers, that the Government is willing to defend and assist British industry.

- Better tax environment: the reduction of the Annual Investment Allowances to £25,000 sharply reduces the incentive for companies to invest in the long-term capabilities of their British plants. Increasing this back to £100,000 or even more would promote commitment to UK production.
- Promote all manufacturers, not select sectors: The Government is too preoccupied with encouraging and funding particular high-tech sectors such as nanotechnology and pharmaceuticals, and even just promoting the retention of R&D work in the UK. This is a dangerous policy, and as various companies have shown, R&D often follows when production is moved abroad. Moreover, without support for all points of the supply chain, and for the low-tech and intermediary manufacturers, there will be little reason for offshorers in these groups to return.

Introduction

Two of the principal 'buzzwords' of the manufacturing industry of the last ten years have been 'outsourcing' and 'offshoring'. It is already well known that many British companies have turned to low-cost nations and pre-eminently to China, to lower their production costs through cheap labour and other boons. In 2010, Chinese exports to the UK were worth $38.8 billion, a rise of 24 per cent on the previous year and making Britain the eighth top export destination for Chinese goods.[1] However this migration trend, which only began in the early 2000s, is slowing down fast and in some sectors, already reversing. Indeed, we are now entering a decade where the new buzzwords for British industry might be 'insourcing' and 'onshoring'. While companies have always moved their production facilities around, this sustained pattern of returning manufacturing to the UK is just beginning. This report examines not only why this will continue, but also why it is likely to accelerate in the coming years. In addition, it will examine the crucial role the British government must play in ensuring onshoring continues.

The offshoring trend has been labelled by critics as a wilful acceleration of the decline of British manufacturing, causing the loss of many jobs in locations already suffering hardship and worsening the UK's trade deficit. The blame for this is said to lie in profit maximisation. The critics are correct, but only to a certain extent. Many companies have had no choice but to offshore production as a means to cope with the increasing competitive pressures globalisation has created, where profits are increasingly narrow. These companies had no choice but to produce their goods abroad, and no doubt would return home if circumstance changed. There is already plenty of evidence to back this up, with many companies repatriating to the UK and shifting their competitive edge from one based on cost to emphasis on quality and service.

Offshorers have found that the business climate in China is turning against them, with many of the advantages, which induced them to relocate there, disappearing fast. For the majority of businesses, this factor is labour costs, and the savings from this are diminishing. Combined with other direct factors such as lower productivity and unreliability, and indirect factors such as rising shipping costs and currency fluctuations, offshoring is losing its sheen. Given this, many companies are basing production abroad to satisfy demand in foreign markets, rather than to import it back to the UK: increasing rather than replacing capacity.

In Britain, while there has been much talk about offshoring and outsourcing, there has been little discussion of this industrial movement back home. In contrast, onshoring has already become the hot economic topic of America, perhaps because it has traditionally valued its manufacturing far more than we have, even if this value has not always been backed up with supportive policies. Industry occupies a special position in the American psyche and the chance to bring it home generates emotional and political reactions rarely seen in the UK, where an industrial decline is often assumed to be inevitable. It is not, and the evidence in this report suggests that many industries, hitherto leaving the UK are now hesitating about doing this. The Government has a second chance to ensure that many stay or return.

This fits in with the Government's concern about 'rebalancing the economy'. It takes this to mean less reliance on financial services and increased development of our industrial capability. Certainly, the rush to cut production costs through emigration undermined the British economy by forcing increased dependence on banking and other financial services to compensate for the loss. On a global scale, onshoring can also be seen as a form of rebalancing of the world economy, where developed economies became overly reliant on low-cost countries for their production and found this unsustainable. Having industrial companies return home and focus on new strengths also offers a large scale internal rebalancing of British manufacturing, away from just acting as the final assembly of products or nominal headquarters, to actually making full products as far as possible in Britain again via domestic supply chains. This will not be winding back the clock to a 'golden age' of manufacturing. Many of the jobs lost will not return simply because the production processes Britain is best at accommodating no longer need as many jobs. Instead, they rely on automated production and as this cannot be avoided, it should be welcomed and incentivised. The jobs these repatriating companies are creating will need new and advanced skills and it is here that the government needs to step in.

The situation

The reports of the death of British manufacturing at the hand of offshoring have been somewhat exaggerated. While smaller British-based engineering and manufacturing firms have roughly halved in number from 1997 to 2010, this is a reduction in capability, not an elimination. Fears about offshoring of manufacturing *en masse* have been raised before: the 1970s and 1980s focused on the domination by Japanese imports and this gave way to concerns about South Korean and Taiwanese supremacy. The shift to China and other emerging economies is the latest worry, and as British manufacturing has survived previous influxes of cheap imports, so it is surviving this wave of moving production overseas as well. As with many trends, the extent of offshoring was somewhat overplayed. In 2009, EEF found that over four fifths of domestic manufacturers only produced goods in the UK while less than ten per cent performed more than half of their production overseas and found that one in seven British companies engaging in offshoring had repatriated their work after finding the promise of cheap and easy production was overly optimistic.[2] For some, the offshoring actually undermined the company's competitive advantage, through unforeseen costs.

As a result of increased global competition, almost all of the businesses that once manufactured high-volume low-quality goods here have either relocated to emerging economies or folded. There was nothing reasonable that could have been done to prevent this, and their loss must be accepted as relatively permanent. China, India and other countries all offered labour at a fraction of the cost of domestic production in the UK and for labour intensive manufacturing, this was a dream come true. Similarly, for uncomplicated high-volume goods, offshoring was a sensible choice. As Figure 1 (p.181) shows, while the UK's export values increased from $285 billion to $472 billion from 2000 to 2011 (or $370 billion to $482 billion at 2011 prices), its global share of exports shrank from 4.4 per cent to 2.6 per cent.[3]

The industries remaining in the UK were those for whom labour costs were not a big factor in total costs: advanced manufacturers with automated processes producing high-value products. This is why some manufacturers, such as paper or glass producers, continued to use UK factories despite the apparent low-tech nature of their products. Cheap labour would not help companies such as these, who employed few people and relied on advanced machinery. Additionally, the bulk nature of these products makes transportation hard, so local production continues to make sense. This is why there is still a surprising volume of 'medium-value' manufacturing continuing in the UK. Strangely, these were also the companies the British government has seemed content to lose, deriding them as producing simple goods unworthy of the UK. Even Vince Cable, Secretary of State for Business, Innovation and Skills, erroneously described how 'metal bashing' had 'mostly gone to Asia now' leaving the UK with just 'high-tech manufacturing'.[4] While this is not representative of the makeup of British manufacturing, it is certainly true that much has been lost abroad.

However, the same international competition that drove offshoring acted at a catalyst for the surviving UK industrial sectors and companies. They have adopted new practices, shifted production methods and rely on the high-quality of their products to win business at home and abroad. Many now work within niche markets, supplying components or services that few others can and relying on their innovative edge to entrench their position within these markets. For the most part, these survivors are now more competitive than ever. The main consequence has been that these companies are producing more with less labour and Figure 2 (p.182) shows how the output of British manufacturing has risen while employment has fallen. It conveys the steady rise in output per unit of labour through automating production, which has been the key to driving down production costs and competing in the global market.

This is not to say that UK companies should be trying to manage their global operations from Britain. There is a clear difference between producing products overseas for overseas markets and manufacturing abroad to import back to the UK. Only the latter is a problem. It would be foolish to think that a business would have the same potential for winning international business if it did not have sales and marketing functions abroad and these must be set up alongside local production for local markets. It would be erroneous to describe this as offshoring as it is a natural by-product of working in global markets and no one expects this to decrease. Indeed, it is an essential ingredient for many companies. While it might be possible to encourage companies to also shift a level of their overseas production for overseas markets back to the UK, this is a secondary concern. Creating a more hospitable business environment for the offshore importers will encourage offshore exporters' migration home as well.

Similarly, many British companies require components produced by other businesses abroad. Since 1994, the importation of intermediate goods to the UK has more than doubled.[5] This is an unsurprising effect of global trade becoming easier and many companies can therefore obtain components no longer produced in the UK. The alternative, of making these goods themselves, will not always be economical. However, it is important to note that the same rule is true the other way round as well: 75 per cent of manufacturers source at least some components from UK companies.[6] However, this international sourcing is also a core problem. If one company in a British supply chain decides to offshore to China, then the likelihood is that it will source its components from Chinese firms, not British ones. As a result, a long string of smaller UK component suppliers are threatened with extinction by this single move.

Defining the terms

The words 'outsource' and 'offshore' are often used interchangeably in the media and while they share some characteristics, they are certainly not the same. A thorough definition of both terms has been provided by the OECD:

> The term 'offshoring' is sometimes used synonymously with the term 'outsourcing'. However, outsourcing means acquiring services from an outside (unaffiliated) company or an offshore supplier. In contrast, a company can source

offshore services from either an unaffiliated foreign company (offshore outsourcing) or by investing in a foreign affiliate (offshore in-house sourcing).[7]

Caldeira UK

Caldeira UK, a Kirby-based manufacturer of cushions, was recently the star of a BBC documentary called 'The town taking on China'. The programme charted the company's trials and tribulations as it attempted to onshore production to Kirby from its Hangzhou factory in China. In the end, the Chinese plant was downsized, retained for low-value cushions and the Kirby factory was prepared to expand, hire more workers and increase the range of products on offer.

Production was first offshored in 2004, at a cost of three quarters of Caldeira's 200-strong UK workforce. This was seen as an inescapable move, as Chinese companies were producing cushions that sold for profit at prices lower than Caldeira's cushions' production cost and Caldeira claimed to produce the cheapest cushions in Europe at the time. The British factory was replaced with 200 Chinese workers working for 20p an hour in a factory five times larger than the retained 50-strong Kirby operation. However, by 2011, wage costs rose to £1 an hour and 150 Chinese staff were let go as a result. Tony Caldeira, the owner of the company, cited rising wages and shipping costs, inflation and the unfavourable foreign exchange rate as the driving forces behind the relocation back to Kirby. However, he could only make the move if the British plant, which produces more expensive, high-quality cushions, could win more orders compared to its Chinese rival, which produces low-cost basic ones.

At a trade fair in Germany, Caldeira found that while the UK factory received fewer orders than the Chinese one, they were bigger, and by more reliable clients. The focus on quality meant the British plant had more potential. Securing large orders gave Caldeira the leverage to invest £50,000 in the expansion of the Kirby plant, to hire more workers and invest in new equipment. However, this investment and disruption had to occur while the order was being manufactured, creating time pressures.

The decision to onshore production to the UK was ultimately made on financial grounds. Caldeira found out that even at the low-cost (and therefore offshorable) end of the market, 'Last year it was 55p cheaper to make a basic cushion in China. With the exchange rate and other costs going up, the difference was only 8p this year.' The conclusion was that production could be brought home now, rather than in the two or three years Caldeira had assumed.

As the definition suggests, once a company has decided to offshore production of a good, it can then decide whether to make the product itself or buy it through outsourcing production to a foreign company. The clarity is blurred somewhat for multinational companies, both British and foreign, who could easily ship components from their plant in one country to another, or from a parent company to a foreign subsidiary. As far as they are concerned, reducing production in Britain is neither outsourcing nor offshoring, but from the perspective of the UK and its balance of trade, it is.

A company's decision about whether to offshore, or offshore and outsource, production overseas is often a strategic one. For businesses offshoring manufacturing, this is a major investment and they would then expect to keep production overseas for many years to mitigate the cost of migration. Alternatively, if the production of components or a service is deemed a secondary activity, this can be outsourced to an independent company on a principal-agent *ad hoc* contract basis. The length of contract can be crucial to the success of the venture: too long, and the potential for being locked-in to a bad supplier are risky; too short, and the delivering firm will not have the motivation to provide a proper, customised service. Outsourcing can apply to the whole production process, where the unlabelled finished good is then imported and delivered to the customer, or just the manufacture of components or intermediate goods, which are then assembled into the finished good in the home country and subsequently labelled. This is not location-specific, and a manufacturer could outsource production to a company down the road just as much to a business in China. For the purposes of this report though, we shall focus only on the outsourcing of production overseas, and to emerging economies in particular. In 2008, two fifths of British manufacturers outsourced the production of at least some of their components to overseas companies, with a further fifth planning likewise.[8]

Onshoring is the mirror image of offshoring, and is the process of bringing formerly British production in foreign countries back to the UK. Where outsourcing has been reversed, and the production has returned in-house, this process is described as 'insourcing' and here, this report is concerned with the insourcing of previously offshored, outsourced manufacturing.

Just as offshoring was driven by 'push' and 'pull' factors, the same is true for onshoring. Originally, it was seen that the cheap labour and new markets in developing countries were significant 'pull' factors for offshoring, while red tape and loss of confidence were factors that 'pushed' companies overseas. For many industries the balance has shifted, and the 'pull' of Britain is beginning to outweigh the attractiveness of China, which has lost many of its appealing aspects. For some firms, this reversal of attraction is not enough on its own to draw their production back home as the time and money invested in offshoring mean they are willing to retain the present *status quo* rather than disrupt production a second time. For these companies, the British government needs to create additional pull factors that will tip the balance in favour of domestic relocation.

What drove offshoring

In terms of imports from China to the UK, there was a steady growth for many years but, as Figure 3 (p.183) shows, this suddenly took off around a decade ago when offshoring began *en masse*. From 2002-11, imports grew four and a half times and the trade deficit with China grew fourfold. This reveals just how recent a trend offshoring is: what might seem an economic inevitability now could not be seen as such on the eve of the millennium. Nor indeed has the pace kept up and 2010-11 has seen a significant slowdown in imports, due in part to the recession curbing demand, but also as manufacturers return to the UK – that year saw the first Anglo-Chinese deficit reduction as exports increased as well.

The attraction of offshoring should not be underestimated. While some sought to take advantage of reduced production costs, other companies began offshoring simply because their rivals were (and therefore assumed this was some sort of key to success), or because their customers were offshoring and they felt they had to move production as well to retain their custom. Entire supply chains were re-localised to service this activity, and it is the relocation of these entities as a whole back to the UK that the Government now needs to encourage.

By 2008, over a fifth of respondents to an EEF survey said they had offshored production to replace UK capacity and another fifth were planning to do so in the future. Over half of the companies had or were going to create offshore plants to increase their capacity.

The excitement around offshoring was palpable in many companies, and a 2004 report from management consultants McKinsey captured this mood well:

> Outsourcing jobs abroad can help keep companies profitable, thereby preserving other US jobs. Those cost savings can be used to lower prices and to offer consumers new and better types of services. By raising productivity, offshoring enables companies to invest more in the next-generation technologies and business ideas that create new jobs... True, some US workers will lose their jobs but this painful reality doesn't weaken the case for free trade… If US companies can't move work abroad, they will become less competitive.[9]

In hindsight, this appears somewhat naïve as it implies these things cannot take place through domestic production and the great offshorer Apple (see p.81) is clear evidence that value for consumers was not a driving force behind their move. Not all were convinced by the value of outsourcing from abroad, as academics Amy Zeng and Christian Rossetti also pointed out in 2004:

> It is generally agreed that manufacturing cost is significantly lower in developing countries; however, the extended distance, the coordination between the partners, and numerous other problems related to international trade often complicated the profit picture... In addition, outsourcing to China involves the increased difficulties associated with differences in culture, language, poor inland transportation and antiquated customs procedures.[10]

Many of these factors have now increased in severity and, in their optimism, many companies simply assumed the cheap wages China offered would last forever.

Nonetheless, the enthusiasm for offshoring also infected the British Labour government, and continues to enthral the Coalition one too. Politicians assumed that British manufacturing would be unable to compete against China and other emerging economies in more industries such as metal goods, toys, plastic consumer products and textiles, and therefore advanced a set of policies that supported financial services and 'new' industrial sectors. By neglecting these older manufacturing areas, scores of firms emigrated and now, without an attractive policy framework in place to encourage their return, it is likely that they, and the hundreds of thousands of skilled and unskilled jobs they provided, are lost for good. This should be a stark warning for the present government.

The tragedy is made even worse when one considers that many of these industries had survived earlier threats from Taiwan, Singapore and Eastern Europe, so *were* able to compete internationally and would have continued to do so, had they not been ignored and penalised politically.

For the most part, both offshoring and outsourcing were driven by similar concerns. The overriding pressure for companies engaging in either is usually the need to minimise production costs and increase profits and viability, often through efficiency gains. EEF's 2009 survey of manufacturers found the first and second most cited motivation behind offshoring were reductions in costs. Almost half of companies moving production overseas identified a reduction in labour costs as a reason for the move and 45 per cent said it was to reduce other costs.[11] Outsourcing in particular was seen as a way for companies to shed the operations that contributed the least to their competitive advantage and to specialise in the business functions that did.

In addition to cost savings, the potential in foreign markets was the third strongest factor driving offshoring, with over 15 per cent of businesses citing this.[12] When utilising offshoring for this, some companies were only offshoring production of goods designed for overseas markets and so supplemented, rather than replaced, their existing British plants. For them, it made much sense to use cheaper production where possible while retaining UK production for Atlantic and European trade. For companies in this situation, Chinese manufacturing was an opportunity, not a threat and in 2008, a third of companies who had not set up production in China identified it as a positive location for business growth and a further fifth expected to set up a presence there by 2013.[13]

On the whole, companies have offshored the least risky functions of their business. In the EEF survey, this is why sales, marketing and distribution services are rated as the most likely function to be offshored, with over 40 per cent of companies expecting an increase in this.[14] By expanding out expeditionary forces, there is minimal risk in entering a new market and companies are not committing many resources. This is not surprising, as it allows companies to see if their product will sell abroad before potentially setting up production there to cater for that market. In this strategy, offshoring tends to complement rather than replace UK production, at least to begin with. An alternative route is to create business alliances with offshore partners and outsource production this way. Should the market prove appealing in a larger way, then firms might consider setting up production bases abroad as well. In comparison though, only 30 per cent of survey companies expected at increase in this. Innovation and R&D, being perceived as the riskiest function, was expected to move abroad by only 15 per cent of companies (although it is noteworthy that this stands at 11 per cent for UK-owned companies and 57 per cent for foreign-owned ones). This move is only sensibly undertaken by companies that have already offshored other divisions successfully. In other words, the first offshoring usually throws up the most problems and expense. In one company, the cost of the first offshore outsourcing was £19,250 for one component, which then fell to around £2,000 for further ones.[15] Once a company begins to offshore, the advantages can grow.

In Britain, reputable recent case studies of the savings made from offshoring are very few and far between beyond anecdotal evidence. One very good study, and the first in a set undertaken by Ken Platts, Ninghua Song and David Bance of the Institute for Manufacturing at the University of Cambridge used the example of a high-tech printer manufacturer.[16] The company used printer cabinets as components, which it began to import from a Chinese supplier that replaced a UK-based one. Looking at the period from October 2004 to November 2005, it examined 51 different variables that would influence the total savings made from outsourcing.[17] Examining the total annual cost of 'price [including carriage, insurance and freight], tax and duty', the cost of purchasing the cabinets from the UK supplier was £1.876 million and £1.488 million from the Chinese supplier.[18] The savings were therefore a not-inconsiderable £388,000, but when taking all other administrative costs into account, the saving was £294,210. However, the Institute noted that this included a one-off currency fluctuation windfall of £27,045, so the total saving from offshoring was £267,165, or 14.2 per cent of purchasing from a British supplier.

Moreover, this case study is a relatively optimistic evaluation of offshoring, as its authors readily admit. The cabinets did not require patent protection and were a mature design not at risk of obsolescence, of which products that change frequently, or have variety, or have changing demand, are at greater risk. They are also uncomplicated, making maintenance of quality easier.

The company already had its own plant in China and was used to finding suppliers and offering support.[19] In short, it is hard to generalise the benefits of offshoring from this example, which even for an industry and time period that favour offshoring, are not breath-taking. Platts and Song followed this research up with a further paper involving five other case-studies. Reading this is recommended, to give the reader a feel for where costs and savings in offshoring lie in different sectors, and how the companies fared.[20]

Benefits of onshoring

For the actual companies onshoring, the advantages are fairly clear. In general terms, they gain greater control over the production processes, and can protect their IP. Joining all the production processes together geographically makes it much easier to monitor quality, identify weak spots and innovate. Depending on the size of the operations, economies of scale could develop, which would drive unit costs down.

On a national scale, onshoring is not just about bringing companies home and increasing UK employment: it has other economic benefits. The US, which is experiencing a similar trend, has seen a decrease in its imports over the last few years as onshoring has taken hold. After a pre-recession peak in 2006 of 11.7 million imported containers being brought to America's shores, this fell to 9.9 million in 2009, a decrease of 18 per cent. The American Institute for Economic Research puts this down to onshoring, rather than the recession, arguing that the trend began prior to this.[21] This means the US is somewhat more self-reliant and, as can be seen in Figure 4 (p. 184), its trade in goods deficit has seen a consequential slowdown that has also reduced the pace of the growing trade deficit as a whole. In 2010, the UK's trade in goods balance was -£98.5 billion and the overall goods and services deficit was -£39.7 billion.[22] Inshoring will reduce this through raising export levels and by increasing the ability to satisfy demand domestically via import substitution.

Even beyond trade balances, there is something to be said for increasing self-reliance. Emerging economies are increasingly looking to move up the value chain and produce advanced goods rather than components or unlabelled finished goods for Western companies. UK businesses should therefore not rely on there always being a ready supply of overseas companies ready to do their bidding. Given the British government and wider EU policy of specialising in high-value and high-tech manufacturing, it makes sense for the UK to be self-sufficient in production of the most critical components required. This is especially valuable given the fast pace of churning out new products in this market, the risk of obsolescence and the rapid pace of technological diffusion. High-quality, reliable component production creates a strong base for companies facing these tough markets to work in.

The exodus overseas also weakened the ability of the UK as a whole to attract business. Demand for raw materials and components in the UK fell as a consequence, leading to further shrinkage down the supply chain. In May 2011, Tata Steel announced a nationwide cutback of 1,500 workers due to a lack of demand, which was only two-thirds the level of 2007 and not expected to recover fully until 2016.[23] Some of these workers have since been re-employed at the blast furnace in Redcar. For many companies seeing part of their supply chain move overseas, the impulse has been to follow suit. With this reversing, it is likely that a new critical mass of manufacturing will grow

The new jobs created by onshoring industry will be no more traditional than the industries they are returning to. As discussed later (see p.110), British manufacturing has changed dramatically in the last few years, and so has the role of the worker, who is no longer making things so much as overseeing machines doing the making. This is not something we should mourn, and while UK production is decreasingly labour-intensive, and manufacturing job creation slows, there is still great value to having this new kind of production in the UK. For a start, the reintegration of British plants into existing supply chains offers opportunities to strengthen the overall manufacturing sector, regardless of there being fewer jobs. In addition, this plays to Britain's existing industrial competitive advantage, which is driven by quality and design rather than labour content. This new work is certainly not to the detriment of the workforce, which continues to find industry a desirable sector as with the need for higher than average skills, employees in manufacturing command much higher salaries than they used to. It pays much better than non-manufacturing sectors: in 2011, mean weekly wages for an employee in manufacturing were £554, compared to a UK mean of £487 and a service industry mean of £474.[24]

Beyond the direct job increases, the Cook Associates report found that in the US, a rise in onshoring would mean a demand for jobs not necessarily directly related to manufacturing:

> [W]e would expect increasing demand for engineering, product development, operations and finance positions... We'll also see demand increase for finance/accounting specialists (CFOs, Controllers) who understand overseas operations and are able to calculate the true costs for exporting since they are difficult to quantify.[25]

These additional jobs should be included under the title of manufacturing employment despite not technically being so, as they are increasingly vital to industrial companies and make up a growing proportion of their staff as actual factory-floor employees wain. More widely, the Bureau of Economic Analysis has attempted to quantify this effect on the overall economy and has calculated that for every increase in $1 of manufacturing GDP, this will create a further $1.42 of economic activity in non-manufacturing sectors.[26]

In addition to direct contribution, onshoring firms are also very likely to engage in activities that boost the innovative capacity the UK needs to retain its competitive advantage. Of all business activity, manufacturing was responsible for 74 per cent of British research and development spending in 2009, despite making up only around 12 per cent of GDP.[27] Beyond formal innovation, and as will be discussed later (see p.90), industrial companies often utilise world-class machinery and rely on sensitivity to customers' needs for business. This requires large capital investment and investment in workers to ensure their skills are up to the job, reinforcing the expertise of the British workforce. This encourages other companies to repatriate production to take advantage of the higher skills as the increased productivity will travel down the British supply chain.

In some cases, whole supply chains can be reinforced through onshoring through the creation of reciprocal relationships. A supplier of components may commission an advanced manufacturing company to create new equipment that streamlines their production process to enhance the product or commission something innovative that adds value to the good. A situation could even emerge where a firm supplies sheet metal to a company to be used in constructing new metal rollers that are then bought by the supplier! This is highly beneficial to everyone involved. While this process could occur between any firms in the global market, extensive research has demonstrated that it was more likely to occur when the relationship between the sectors was in close geographical proximity.[28]

There are also strong environmental benefits to onshoring industry. The environmental cost of shipping goods from countries in the Far East is a high one, but can be avoided altogether by sourcing components domestically. Moreover, companies within the European Union have far stricter Directives and regulations that they must follow with regard to carbon emissions, so if the EU really wants to reduce global emissions it should ensure Low and Medium-tech (LMT) companies are retained within their sphere of influence to ensure no 'carbon leakage'. This has already been seen in some industries such as cement manufacturing, where much of it is now made by European firms in extra-EU locations such as northern Africa, where emissions regulations are laxer. The products, made in more polluting processes, are then imported back to Europe. Britain has therefore lost its cement manufacturing sector while global emissions are made worse, not better. Overall, the UK still has some of the most world's most energy and emission efficient manufacturing, even within energy-intensive sectors, and onshoring more businesses ensures further green production.

The Bedlam Cube

In September 2007, British toymaker, Bedlam Puzzles, announced its decisions to repatriate its production from China. For this purpose, founder Danny Bamping (who in 2005 famously turned down financing from the BBC's Dragons' Den for the marketing of the firm's main product, the Bedlam cube, now known as the Crazee Cube) set up a new company, UK Manufactured, in conjunction with P&P Holdings, the Wiltshire-based plastic injection moulding firm.

The toy industry, which sources more than two thirds of its products from China, provides a telling example of the problems of product quality and safety which can all too often accompany offshoring. Although often cost-effective, outsourcing and, especially, offshoring increase the complexity of the network in which companies interact, increasing the likelihood of miscommunications and inaccuracies during the production process.

According to Mr Bamping, the most intractable problem he encountered whilst working with Chinese suppliers was that of unpredictable quality: 'certain orders did go wrong. I then had to remake the cubes quickly and airship them in from China, which is hugely expensive – not just financially but also environmentally'. Although the Bedlam cube costs £3 to make in the UK, compared to just £1 in China, he believes that this cost is more than compensated for by the better control of product quality afforded by manufacturing at home.

It seems that the public may agree with Mr Bamping on this point. The entrepreneur ran an experiment in the run up to Christmas 2007, lowering the price of the Bedlam cube by £1 to £9 on the company website for those cubes still being made in China at the time, whilst continuing to sell the British-made cubes for £10 each. The website (which only shipped to the UK) sold five British made cubes for every cube produced in China. Of course, it is not possible to accurately assess the motivations of those consumers who opted to buy the more expensive, British-made cube, but it seems likely that expectations of quality, as well as patriotism, came into play.

The focus

This report is focused on the latest wave of repatriating manufacturing from emerging economies, and China specifically. It will not be a discussion about the service sector. While outsourcing has certainly occurred in this area, and allusions to the stereotypical 'Indian call centre' continue to be rife, the British service sector has not been adversely effected by this migration. A comprehensive International Monetary Fund study of service outsourcing in 2004 found that while the UK is one of the top importers of services in absolute terms, if countries' importation of business services is scaled to GDP, the UK was 85th in the world with slightly over one per cent of GDP outsourced.[29] The largest importers on this scale were small economies such as Angola and Ireland. Moreover, the UK was the second largest beneficiary of global outsourcing, with the UK gaining $27 billion this way. The conclusion was that 'jobs displaced by service outsourcing are likely to be offset by new jobs created in the same sector'.[30] There is a level of onshoring within the service sectors, and from 2008 to 2009, India's telecommunications exports fell by 40 per cent. While this was partially caused by the recession, the trend has continued, with a further 5 per cent shrinkage in 2010 and there are plenty of instances of call centres being relocated to the UK such as by Santander and BT, often in response to customer pressure.[31]

The focus on British companies' experiences of China is a reflection of the offshoring trend, which has seen the lion's share of manufacturers relocate there. China is also seen as the greatest threat and opportunity for British businesses, beyond all other low-cost countries. In 2009, a survey of manufacturers found China to be the top challenger for businesses, cited by almost half of respondents. In terms of opportunities, it was top non-Western destination for companies with around 30 per cent for companies saying there were opportunities there.[32] For British companies offshoring to take advantage of cheap labour, China was the destination of choice for half, with India being chosen by another third. Figure 1 (p.181) demonstrates the speed with which China rapidly grew its share of global exports, from 3.9 per cent in 2000 to 10.4 per cent in 2010, the value of which increased from $250 billion to $1.6 trillion, a rise of 500 per cent.[33] This was far beyond any other country and reinforced China as the pre-eminent offshoring location.

India is certainly an attractive location for offshoring British business, and Figure 1 (p.181) shows that from 2000 to 2010, India saw its global share of exports increase from 0.7 per cent to 1.5 per cent, the value of which rose from $42 billion to $221 billion.[34] However, India is primarily a location for service offshoring. It is now the seventh largest provider of services worldwide, producing 3.3 per cent of global service exports in 2010, with a growth rate of 33 per cent which was faster even than China's.[35] In terms of manufacturing offshoring though, India pales into insignificance with China and only 15 per cent of companies identified India as the place they would choose to expand their business abroad.[36] This is unsurprising. China currently has 215 million citizens working in industry: 58 per cent more than the entire manufacturing workforce of the rest of Southeast Asia and India combined.[37] Figure 1 (p.181) reinforces this point and while the term 'the BRIC nations' is bandied about, it is clear that India and Brazil and Russia are small-fry compared to China in terms of global exports.

As China and India become more expensive, it is expected that some companies will shift production to other countries offering cheaper labour and some countries already offer very competitive wages indeed. At present, hourly compensation is $1.80 in Thailand, 49 cents in Vietnam, 38 cents in Indonesia and 35 cents in Cambodia.[38] However, the number of companies making this move will be comparatively small compared to the number who made the original exodus to China. Despite the lure of even lower wages, these countries cannot offer anywhere near the scale of resources most companies will need to make offshoring worthwhile. Human capital is not available in the volume required, especially by larger companies, and worker skills are wanting, meaning productivity is lower. Infrastructure is also precarious, making delivery and supply chain integration difficult and unreliable. There are additional concerns such as corruption that mean all in all, these are not viable destinations for many businesses. Analysis of the US market found that of American industry migrating from China, 75 per cent would go to the US with only the small remainder going to other low-cost countries such as Mexico.[39] For the majority of British companies in the same situation, they will either keep their production offshored in China or repatriate it to the UK. This is not likely to change dramatically for many years to come.

The report will not examine offshoring to Europe or other developed economies because it is harder to generalise about the reasons why movement there might take place: it is far more down to individual companies' circumstances. Nonetheless, given its proximity to the UK, Central and Eastern Europe offer British manufacturers a very attractive trade-off between low wages and logistics and have been key locations to which the UK has lost activity. It is still much cheaper than the UK to establish production in Central and Eastern Europe (though more expensive than in Asia) and there is much less distance for goods to travel. This is a powerful combination and it was for this reason that in a 2008 survey, while 20 per cent of businesses seeing opportunities for growth abroad identified China as the place they expected to set up new operations, 30 per cent chose Eastern Europe.[40] The relatively high skill and productivity levels, combined with cheaper shipping costs means this area is attractive for bulky, high-tech goods production and is a favourite with the chemical, machinery and transport equipment industries. At the same time, Britain has also seen some onshoring even from these locations. Jaguar Land Rover has switched the supplier of some of its automatic gearbox cooling pipes from a firm in Turkey to British company Lander. Lander, which employs 280 workers in the Midlands, won this contract through offering higher quality products at a competitive cost.

1

What drives onshoring?

Quality and reliability of offshored production

The label 'made in China' has never been viewed in a positive way and for years it has been frequently used as a metaphor for goods that are cheaply made and of poor quality. Another apt saying would be that there is 'no smoke without fire', and it is clear that many UK companies have a great deal of trouble with quality when they offshore production of their goods to China. For instance, Kevin Steers, the CEO of the British barbed wire manufacturer Betafence, shifted production of barbed wire to Tianjin in China. He found: 'the quality was poor and the service was terrible' and duly resumed production in Sheffield.[1] Interestingly, the company had even shipped the machines that made its barbed wire out to the Tianjin plant, so in theory, there should have been no difference in quality, only in the cost of production. Whether orders were lost in translation or simply through a lack of care, many British companies have experienced the same outcome.

Some companies have found their ability to offer products severely constrained by offshore outsourcing: their foreign partners are happy to make a particular type of product but only in a particular way. Sometimes, this is only found out after production has already begun and when alterations are required, the partner is unwilling to implement these. This is not ideal for the offshorer, who can find that they have to restrict their own product range, or are unable to deliver the quality of customisation to wihch their customers are used. For some, this could be a deal breaker and business might be lost. While this does not matter so much for large companies working in huge bulks or those with brand names that automatically sell the product, flexibility of production is still very important for the majority of manufacturers and without this, they risk falling behind market trends. At a basic level, the problem with offshoring for many companies is one of reliability. Companies cannot rely on their Chinese suppliers to deliver satisfactory products. Indeed, this was the most mentioned problem by far in EEF's 2009 survey of offshore outsourcers, with 38 per cent of small, 55 per cent of medium and 53 per cent of large companies all complaining about quality problems.[2] Reliability issues breed distrust and often necessitate a full examination of the production processes. In one case-study from Platts and Song, a British manufacturer of control panels with production outsourced to China had to reject shipment of components because the quality was not high enough. Alongside rigorously checking the next batches coming through, the managing director of the firm had to visit the Chinese supplier to find the source of the issue. It turned out that the supplier had been using low-quality, recycled copper rather than the necessary grade required.[3]

Often, when the quality is so poor that the goods cannot even be shipped back to China to improve them, they simply have to be discarded. In this case, the Chinese supplier often only refunds the cost of the product, and not that of the shipping and any other indirect cost, which are simply counted as losses. In the case of a British tape measure manufacturer, the batch it received from China was so unsatisfactory that the supplier gave them a refund, but:

> The supplier did not pay for the indirect loss including transportation, receiving and inspection cost. In addition, before the faulty products were thrown away, it took 3 workers one day to chop off the ends of the tape, in order to avoid other people picking them up and selling them.[4]

Tranquil PC

The reversal of Tranquil PC's offshoring strategy began with the purchase of a £50,000 CNC milling machine last summer, an investment which was supported by funding from Lombard, the assert finance division of the Royal Bank of Scotland. Founded by David Thompson in 2003, Tranquil PC has made a name for itself as a designer and manufacturer of computers which use low amounts of energy and are nearly silent. It has been able to get rid of noisy fans by designing aluminium computers cases in such a way as to naturally dissipate heat from the internal electronics. Its silent systems can run digital TV playback, CD, DVD, Blu-Ray and digital audio. Its newest home media centre, the ixLS, is rapidly gaining a reputation as one of the world's leading media centres.

According to Mr Thompson, Tranquil PC moved offshore because:

> At the very beginning we would get aluminium from an Italian company, steel cabinets from a local company in Lancashire, and then we would assemble those products. As volumes grew, those suppliers couldn't keep up with demand so in 2004 we moved the creation of our cases to China.

Instead of the cost-effective aluminium extrusions being produced in China, Tranquil PC is introducing aluminium block milling, a process whereby a product's 'skin' is carved out and then a pocket made in it. By the end of this year Mr Thompson plans to have at least another three CNC milling machines operational at the company's Trafford Park premises.

This move towards developing an in-house manufacturing capability has been driven primarily by the need to find a balance between keeping costs low and flexibility. Over time, frustrations with the inflexibilities imposed by their Chinese supplier grew:

> We were limited to the number of variations that the Chinese supplier could make for us... If we bring in a new variation we have tooling costs, we have minimum order quantities and, once we receive those cases after eight or 16 weeks, we then have to sell them. We can't change them, we can't change the colour, shape, size, features, function or anything else. That is not a problem if you don't want to expand your business, but we do.

Increased flexibility of production will enable Tranquil PC to offer its growing customer base - which includes the BBC, Alcatel Lucent and T-Mobile - a tailored and more efficient service.

Even when quality eventually improves to the required standard, this can take time and numerous batches of unsuitable products. This is a short-term inconvenience, but given the distance and length of time before delivery, the problem can continue for a long time.

In addition, issues relating to the unreliability of delivery were cited by 18 per cent of small, 30 per cent of medium and 28 per cent of large businesses as major problems they face because of offshoring. Steers of Betafence found: 'we would put in orders in January and February and they would come in one big shipment in May. When we make it ourselves here the turnaround is a few days'.[5] The issue here is clear: when outsourcing, companies are unable to directly control of production and they also lose control of the delivery process. For a myriad of reasons, the result is often chaos. While this might annoy customers of Betafence, for companies outsourcing the production of components, this has a serious knock-on effect through the supply chain and the longer the chain, the increased likelihood of disruption and of this being damaging. Those previously working along the principles of 'just-in-time' manufacturing have found this impossible given the inability to fine tune their orders by knowing component delivery dates. Some companies have found that to receive orders on time, they must be placed many months in advance. This is acceptable for companies producing goods at a continuous pace, but that is a rare company indeed. Most will alter production according to demand, and this is not something that can be easily foretold almost a year ahead, making the ordering of too many or too few components a real risk that then cannot be resolved easily. As discussed later (see. p.102), the reliability issue has led some companies to onshore to the UK as a means to make the supply chain more resilient.

One way of circumventing these problems is to ensure the risk of them occurring is minimised as far as possible. This can be done (especially for quality issues) by sending British managers out to China to supervise production, but if this is a long-term arrangement, it in turn creates more costs. The University of Cambridge found that to send a manager abroad, it costs roughly double the salary that would otherwise be paid to them if they did not relocate.[6] There are addition costs such as airfares home, insurance and school fees if the manager has a family. Obviously, this also means these competent managers are no longer available in the UK, so the company could discover complications at home stemming from this. Ultimately though, there is no way round the delivery times issue.

Wage inflation

The initial movement of manufacturing abroad was most frequently due to the cheap labour on offer and wages were so low that almost any other faults were forgivable. This is no longer the case. The average compensation of Chinese manufacturing labourers has steadily risen and in six years, from 2002-08, it exactly doubled. This data takes into account direct pay, social insurance expenditures, and labor-related taxes. Of course, the end result is that a Chinese worker's compensation cost is still a small proportion of that of a British employee. It will

Trunki

Rob Law first came up with the concept of a children's ride-on suitcase in 1997. In 2006 his invention, which he named 'Trunki', was premiered on the BBC's Dragons' Den. By 2011, 300,000 Trunki's had been sold in 62 countries. Last year Magmatic, the Bristol company set up by Mr Law, turned over £6 million and cleared a profit of £1 million.

At the beginning of this year, Magmatic employed just 23 people at it Bristol headquarters and every Trunki sold in the UK was manufactured in China. From mid-April 2012 the company began bringing home its production to the UK. By the end of this year every Trunki sold in Britain will be designed in Bristol and made in England, with the first British-made version a licenced Team GB and London 2012 Trunki.

According to Mr Law, although it is more expensive to manufacture in the UK than in China, other benefits outweigh cost, most notably the significant advantages conferred by shorter lead times. Inject Plastics of Totnes, Devon will be taking over production, in so doing reducing order lead times from 120 to just 30 days. For customers, this means quicker and more reliable delivery. It will enable Magmatic's sales team to react more quickly to market demands and promise customers that Trunki will be 'NOOS' (never out of stock). Shorter lead times also facilitate design innovation - as does the absence of language barriers and time differences - which, Mr Law argues, enables the creation of 'really innovative and well-engineered products'.

not catch up for a long time. In 2008, the last year for which clear data is available for both the UK and China, the cost of a Chinese labourer ($1.36) was only 4 per cent of that of a British one ($34.22).

While Chinese workers are still cheaper than their UK counterparts, it is clear that the boom years of ultra-low cost labour are over and the cost of the formers compared to the latter is rising fast. While the rate of yearly growth in UK wages from 2002-08 was 9.1 per cent, growth in China was 33 per cent per annum. However, since the recession, while a data breakdown is unavailable for China, manufacturing compensation in the UK shrunk to $29.44 per hour by 2010.[7] This decline means a yearly wage increase from 2002-10 of just 4.1 per cent, which is much slower. According to the Boston Consulting Group, which took the effect of the recession into account, over the period of 2005-10, Chinese wages grew by 19 per cent per annum. In comparison, the UK's wages actually shrank by 0.2 per cent over this period.[8]

The national and regional Chinese governments are surprisingly forceful in pushing up minimum wages. In 2012 alone, Beijing raised its minimum monthly wage in January by 8.6 per cent to 1,260 yuan. In February, Shenzhen, the manufacturing centre of southern China and one of the most desirable locations for the offshorers, raised its minimum wage by 14 per cent to 1,500 yuan. The coastal port city of Tianjin had a 13 per cent hike to 1,310 yuan starting in April.[9] These are only a few of the rises and in total, 20 regions have experienced minimum wage hikes of over 20 per cent, a growth with no foreseeable end in sight.[10] Raising the national income level is a key aim of the latest Chinese Five-Year Plan, something of which offshorers will be wary.

There is no reason to suppose that rising labour costs in China will fall in the future, and it is this trend, that will affect the decisions of manufacturers contemplating onshoring, more than the actual wage costs at present. Boston Consulting Group predicts the average Chinese hourly wage to rise to $4.51 by 2015. In the Yangtze River Delta, the centre for offshored high-skill manufacturing and the region with the highest industrial output, it expects wages to reach $6.31 per hour by that year.[11] Assuming a continued (and generous) average growth in UK wages by 4.1 per cent, this would mean average British manufacturing wages of $36 in 2015. At a sixth of UK wage packets in the Delta, and an eighth elsewhere, up from a twenty-fifth in just seven years, it is hard to see China as the continued destination of offshoring.

While national wages are a useful tool for macro-analysis, it is also the case that companies' offshore production is normally located in the most expensive areas of developing countries. In China, the vast majority of foreign factories are in the economically developed provinces around the Yangtze River Delta and cities such as Beijing, Tianjin and Guangdong, where wages are at a ten to fifteen per cent premium.[12] Relocating to the UK though, often means setting up shop in traditional manufacturing heartlands that have suffered in recent years and therefore have lower average wage rates. Median full-time gross hourly earnings in the North East of England are £10.31 and in Yorkshire and the Humber are £10.32. This is nearly ten per cent lower compared to a

UK average of £11.20. In the South East and London, both predominantly service-focussed areas, hourly earnings are £11.81 and £15.67 respectively.[13] Of course, China too has areas where manufacturing wages are still low, but these are mostly the rural areas that lack the necessary infrastructure. This compromise is one unlikely to attract many offshorers. The American Institute for Economic Research calculated a comparison between the higher wages commanded in the industrially developed parts of China and low wage American states and found that: 'productivity-adjusted labour costs in Shanghai will be 30 per cent of that in Kansas'.[14] This difference would be very similar if Teesside was substituted for Kansas, although this also depends on the sector.

Beyond general rises, there have been high profile cases of companies being suddenly hit by wage hikes. In 2010, the Taiwanese company Hon Hai Precision Industry Co., Ltd, which owns the subsidiary and more well-known Apple product assembler Foxconn, raised its average monthly wages at its Shenzhen factory from $132 to $294.[15] In February 2012, it raised them again by up to a quarter in response to a spate of employee suicides and negative press, putting entry-level pay at $185, double that of three years previously, which increases further after a three-month probationary period.[16] Industrial action has also been increasingly voracious, and in 2010 Honda was forced to raise its wages by 24 per cent at its Chinese factory to satisfy striking workers, losing $3 billion in sales during the strike.[17] While this only affects individual companies initially, Chinese workers are increasingly able to access information about their compatriots' salaries via social networking and the internet. In an increasing number of cases, this has meant that they leave existing jobs for better pay elsewhere, or threaten to leave them unless they also receive more money. This force is pressurising a universal wage rise.

Beyond actual wages, companies are finding there are extra expenses to pay for their workforce that were not required ten years ago. In previous years, many factory workers were agrarian migrants, who worked there because of the attractive wages and sent a regular portion of their salary home. Given they were earning significantly more in the factories than in the fields, they were content with basic provisions of company welfare: the job was a means to an end. Now, the new influx of workers demands more from their workplaces. Silverlit, a toy manufacturer from the Pearl River Delta employs 5,000 people. Its deputy chief executive, Eddie Wong said: 'Unlike their parents, who only wanted to work hard to send money home, these young workers also care about quality of life'.[18] While unsurprising, this means Silverlit, like increasing numbers of companies has to pay more than minimum wage and provide extra-curricular facilities such as recreation grounds and libraries. Given the comparatively luxurious treatment of Western manufacturing employees, it might appear callous to label these as sunken costs or unproductive capital investments. It goes without saying that all workers should be able to work in safe, clean conditions and gain fulfilment from their work. However, from the perspective of the offshorers, these are costs that did not previously exist and they have to be taken into account as additional production outlays. A more enlightened employer might see that a happy workforce is a more productive workforce, but the improvements delivered this way might also be considered too marginal to justify the cost.

Even including these extra costs, it is clear that the labour cost savings made from offshoring production to China are still fairly high. It is important to remember though that labour costs actually constitute an increasingly small proportion of total production costs, making the savings marginal. In British manufacturing, the labour content of goods varies considerably by sector, from around ten to forty per cent but in 2009, the latest year for which data is available, wages only made up 22 per cent of average total production costs in British manufacturing.[19] Given that, the wage savings of retaining production in China by 2015 are actually only 18 per cent on the east coast and 19 per cent inland, which is a much less impressive saving than commonly assumed. Given the general trend towards a declining labour content, this figure will only shrink in time. Beyond labour, there are numerous other costs to be taken into account, which are described in the various sections below, but their aggregates can completely swamp labour savings. One British company, Bedlam Puzzles (see p.22), used to make all its products in China until 2007, when costs had increased by around 20 per cent, excluding other factors such as shipping costs and delays.[20] Accounting for all these other costs, total cost issues were the second most cited complaint by UK companies with overseas suppliers with negative responses from 32 per cent of small, 33 per cent of medium and 40 per cent of large suppliers.[21] Reflecting this, Joerg Wuttke, of the EU Chamber of Commerce in China has predicted 'that the cost to manufacture in

China could soar twofold or even threefold by 2020'.[22]

UKLED

UKLED is a Wirral-based manufacturer and installer of LED lighting systems. The firm, which describes itself as 'small but innovative', focuses on the 're-lamping' market - that is, the replacement of bulbs and tubes in existing commercial lighting systems with LED equivalents. Directors Mike Parker, Colin Griffiths and Tony Griffiths claim that this process can reduce customers' annual energy bills by up to 70 per cent.

For a year prior to December 2011, UKLED had offshored the manufacturing of its lighting to Chinese producers, mainly because of the traditional association of China with cheap labour. However, rapidly rising Chinese wages prompted the company's directors to think about relocating its manufacturing to the UK; the total cost advantages of producing in China decreased significantly when additional costs associated with offshoring, such as shipping, were considered. Repatriating the production of the company's LED equivalent to florescent tubes was facilitated by the installation of more sophisticated equipment at its base in Bromborough; this new technology decreased the labour content of the production process, allowing UKLED to set up a low-cost manufacturing operation in the UK. The overall cost of the move was £80,000, a quarter of which was provided by a grant from Wirral Council's Business Support Team.

UKLED soon found that repatriating production conferred additional benefits, namely higher quality products and far fewer delays in production and delivery. Colin Griffiths, Technical Director of UKLED, commented on the move:

> We are the only UK manufacturer of LED lighting systems and this is already proving to be a major advantage. We feel that the next natural step for us is to supply direct to contractor and specifiers who are able to benefit from a high quality, guaranteed, competitively priced LED product that is manufactured in the UK.

Declining access to workers

Companies are increasingly finding that there is a shortage of incoming skilled labour in China, and this has been fairly instrumental in pushing up wage costs as well. In such a populous country, a labour shortage might come as a surprise, but given that the majority of the skilled workforce has already been absorbed into industry, this means the incoming supply is primarily made up of new graduates. The majority of these new graduates will have to be trained to use industrial machinery, which means investing money and time. This is unattractive to most offshoring companies, who want a ready prepared workforce.

On top of this, demand continues to be greater than supply, a trend that will be aggravated as more Chinese companies begin producing goods for the domestic market, stretching the availability of workers for offshorers. For example, when Foxconn began recruiting employees for a new plant opening in Hunan province, it anticipated that it would hire 3,000 skilled employees but found it could only muster 100.[23] This issue increasingly affects the entire eastern seaboard of China. In March 2011, the China Daily newspaper reported that in the province of Guangdong, where offshorer Mecca Shenzhen is located: 'private companies, will suffer a shortage of more than one million workers this year. Guangzhou [the capital of the province] alone is expected to see a labor shortage of 150,000'.[24] As a result, minimum salaries in Guangzhou were raised to 1,300 yuan (£130), an increase of 18.3 per cent on its previous level and the largest rise in mainland salaries in 2011. The aim of the hike was to attract more migrant labour from the central Chinese heartlands.

However, migrant labour is not as voluminous as it once was. China's own development has seen more work available in central and western provinces and wages there have also increased. For many migrants, these new opportunities and the ability to remain near home outweigh the benefit of a marginally higher salary in a faraway province. Companies often find that migrant labourers return to their home village during the Chinese New Year holiday and one offshorer, a pram manufacturer, found that only 85 per cent of his workers returned after the break, compared to 95 per cent the previous year.[25] For an industrial town such as Shenzhen, whose 12 million population includes 6 million migrant workers, such a loss will significantly affect recruitment ability and attractiveness.[26]

In addition, the rapid rise in industrial wages means many potential employees can shop around for work, and effectively bargain for higher salaries. Tony Caldeira, of Caldeira UK, (see p.8) said: 'people view the Chinese workforce as docile and subservient, but this new generation has more confidence and is almost cocky'.[27] Caldeira found that his employees rebelled at the offer of a 30 per cent pay rise and wanted 50 per cent or they would leave. Even at Chinese job fairs, companies have been known to up their salary packages by 50 per cent during the course of the day, as a means to win over the limited workforce available. Even after accepting the offer of an interview, potential employees go to inspect the plant before committing, to examine the conditions compared to rivals. Caldeira concluded: 'In Britain, it's the factory who picks the staff, in China, it's the staff who pick the factory.'[28] The unemployment situation in old manufacturing areas in Merseyside is such that when recruiting for unskilled labourers, 80 people turned up for Caldeira, hoping for an interview, even when offered just minimum wage.

While a lack of unskilled labourers is not so pressing, the universal demand for skilled workers has meant that the wage difference between the inland and coastal regions still seen for entry-level jobs no longer exists for those with higher qualifications. When one company examined the prospects of relocating to Hubei from the coast, it found that its costs would be just five to ten per cent lower and this marginal saving would likely be negated by the higher transport costs incurred.[29]

China's demographics also work against it. The one-child policy has meant fewer numbers of incoming entrants to the workforce and will continue to act as stranglehold for almost two decades, even if it were revoked tomorrow. The slowdown in the population growth rate has been steady since 1988's rate of 1.61 per cent but has been most noticeable from 1998 to 2009, when it fell from 0.98 per cent to 0.51 per cent.[30] Neither situation works in favour of offshorers. Rising skill levels mean workers can demand higher wages, and if this was restricted, the bottleneck would cause productivity issues. In the long run, it is simply unsustainable for China to continue to be a paradise for cheap labour intensive manufacturing.

Productivity

Chinese workers have a reputation for working very long hours, living on the factory site and taking only one holiday a year. These would appear to be the traits of a hardworking workforce, but some offshorers have found these conditions the necessary trade-off for the lower productivity of their workforce. Indeed, productivity has always been an Achilles' heel for offshoring companies who often find that while labour is cheap in developing countries, it is not very efficient. Tony Caldeira, pitting British and Chinese workforces against each other found that productivity in the UK was much higher. He said: 'my UK staff seem to have a longer attention span and they are able to focus for longer periods of time, whereas in China, they tend to work for longer hours but don't tend to produce as many products per hour.'[31] On the one hand, there is a great deal of potential to increase productivity, through automation of production or upskilling the workforce. On the other hand, this normally requires intensive investment, which undermines the reasoning behind offshoring in the first place. It is therefore a major determinant of competitive advantage and a critical factor in the onshoring assessment.

China, like other emerging economies, has low productivity at present and therefore has the highest productivity growth rate in the world as measured by GDP per head in purchasing power parity terms, which is growing by 12.5 per cent annually.[32] This is because unlike the UK, the 'low-hanging fruits' of easy, cheap productivity increases are still possible. All these countries need to do is import the existing technologies already made available in the developed world. For the UK to continue to increase its productivity, new technology must be developed and this requires both time and investment. Little wonder then that from 2001-10, China's labour productivity grew by 9.5 per cent, the fastest worldwide while the UK's rose by just 0.9 per cent.[33] Nonetheless, this impressive rate of growth belies the fact that actual productivity in BRIC nations is still very much lower than in the UK. In 2009, the UK's productivity was ranked eighth in the world.[34]

In 2010, China's manufacturing output was estimated to reach $1.923 trillion, compared to $231 billion in the UK. However, it took 100 million Chinese workers to reach this, compared to 2.9 million in the UK. This would imply British workers are four times more productive than their Chinese counterparts and even measuring by manufacturing gross value added, the UK produced $3,717 per employee, compared to $1,459 per Chinese worker.[35] As with labour costs, while Chinese productivity development is fast, it will be a while yet before theirs is anywhere near rivalling the UK's.

However, further increases in Chinese productivity are not inevitable. In rapidly growing industries such as smart phone assembly, the sheer growth of these sectors has stalled productivity rises: the continual need to hire more staff to satisfy demand and produce increasingly complicated products, without investing in further automated equipment and training, has dragged down productivity and value per employee dramatically. From 2005 to 2010, Hon Hai's subsidiary, Foxconn International Holdings, more than doubled its workforce of product assemblers from 59,000 to 126,500. Over the same period, it saw its productivity fall from $108,000 to $52,000 per employee, and value added per employee from $13,750 to $6,300.[36] This rapid expansion was a fairly damaging manoeuvre, which steadily whittled away at the company's pre-tax profits, from $419 million in 2005 to a loss of -$176 million in 2010.[37] In part, this was because the increased demand from smartphone manufacturers meant Foxconn had to pull out of non-smartphone contracts that had proved lucrative and a lifeline. For its customers, this is a concerning trend, and few could risk their supplier going bankrupt, but at the same time would not want to constrain production for this reason. While secondary suppliers could always be found, dividing the workload places a limit on the economy of scale advantage that a single large dedicated factory can provide. For the Chinese company itself, increasingly narrow margins meant Foxconn had to cut back on 'non-critical' costs elsewhere, and according to the Centre for

Research on Socio-Cultural Change (CRESC), it '...created the anomic conditions under which suicides proliferated... responsible health and safety procedures also became difficult to operate in an environment of squeezed margins and intensification.'[38] All this led to a period where Foxconn's poor working conditions were in the media spotlight throughout late 2011 to early 2012, which reflected badly on Apple, its primary customer. It was also these issues that led to the wage hikes discussed earlier (see p.37). In addition, Foxconn offshored some production itself to Vietnam and Indonesia, and built new plants in central China.

Chinese productivity growth has been driven by intensive foreign investment, which has increased the number of machines and computers in factories and therefore helped make the average worker more productive. These are easy pickings to make for a national rise in productivity, as is the influx of rural workers to cities. They leave behind agricultural jobs with low-productivity for the more productive manufacturing ones, raising the national average level without a need for much investment. However, it should be remembered that while there are plenty of ways to improve productivity further, such as through increased automation of processes, the fact is that this requires investment on a scale companies attracted to cheap labour will be unwilling to make. In the Caldeira cushion factory in China, the cushions are still stuffed by hand, whereas in the UK, there are machines that automatically stuff the cushions and recycle any wasted stuffing.[39] Automating production means the labour content of products goes down, as do labour costs as a proportion of total costs. In other words, a productivity drive in China will undermine the country's existing primary competitive advantage.

Even when companies have engaged in heavy investment to maximise productivity while offshoring, some have already brought production home. The British manufacturer of flooring, Amtico International, did exactly this, once it learnt how Chinese factories innovated to increase efficiency. It combined these improvements with further advances made at the British end to create a hybrid production process that made UK production more feasible. On top of this, the company found that, while it had been 30 per cent cheaper to manufacture in China initially, its costs were going up by eight to ten per cent per annum. It decided to onshore back to Coventry, creating an extra 100 jobs.[40]

New Balance

US-owned New Balance opened its first factory in Cumbria in 1982, employing 40 people and taking over the premises that Clarks Shoes had closed down. The athletic apparel and footwear company moved production to a larger factory in Flimby in 1991, where it continues to design and manufacture its products today (despite sourcing their materials from the continent). New Balance makes roughly one million pairs of 'performance' shoes and 225,000 pairs of 'lifestyle' shoes every year. Several of the factory's 250 employees have been working there for almost 30 years.

When asked why New Balance has chosen to maintain it UK factory rather than offshoring production, factory manager Andy Okolowicz emphasises craftsmanship: 'You could import some of the products from China but that's not what it's about. We have years of experience.' However, Okolowicz does argue the case for government investment here, stating that 'the amount of training we have to do to keep our skills up is expensive.'

He believes that factory's workforce provides not only unmatched skill levels but also impressive productivity and flexibility. 'They realise it's tough out there, that we're in it together, and they feel safe here because we're not in it for the quick win.' They have, he argues, worked 'incredibly hard as a factory' to compete with low cost overseas manufacturing and have introduced new technology such as robotics. As a result of these investments, productivity has improved by 35 per cent, and the factory has cut lead times from 'about three weeks to three days. Our ultimate aim is to cut, stitch, and make a pair of shoes in less than a day.'

Regulation and intellectual property infringement

Many offshoring companies have underestimated the increasing complexity of Chinese law and continue to be unsatisfied with it. One the one hand, they experience intellectual property fraud with Chinese companies counterfeiting their goods but with little restitution, and on the other, companies are required to operate in increasingly strict conditions. While this is arguably beneficial to Chinese workers and society, it leaves offshorers in the cold.

On a national scale, an overhaul of employment law occurred in 2007, with a new set of rules called the New Labor Contract Law (NLCL) effective of 1st January 2008. This was designed to protect workers, but did China considerable economic damage during the recession that began later that year. The NLCL implemented a 40-hour working week, with only 36 hours of overtime allowed per month: this is more constraining than in the UK, where for the most part, any length of overtime is acceptable provided the employee consents to it. It basically means companies in China have to employ more people to do the same job and workers themselves are unhappy at the inability to supplement their usual wage as much as they need to. Additional regulations are now in force, granting workers more rights. The NLCL requires employers to pay social insurance to employees and in addition, when a company lays-off staff, it has to pay one month's salary in severance pay for every year that they worked there, to a maximum of twelve months. Unlawful terminations, i.e. before the end of a contract period, are paid for at twice this rate. Furthermore, the beginnings of union representation in companies was enshrined, with workers granted the right to politically participate in the business's operations. For offshorers, whose interests are not normally linked to the improvement of social conditions in China, this can create fraught relationships, disagreement and undermine productivity. These new rules are not just symbolic and are clearly being enforced by the State and utilised by disgruntled employees. The Chinese Ministry of Human Resources and Social Security has recorded that in 2008, there were 693,000 cases of labour disputes which was double that of 2007, the year before the laws were introduced.[41]

In response to a survey about the effects of the NLCL, a third of companies reported that it had pushed up average labour costs by over 30 per cent, and the other two-thirds all claimed costs had risen between 20 to 30 per cent. Despite the need to hire more workers to comply with the overtime regulation, the NLCL was also felt by two-thirds of respondents to make hiring 'much more difficult'.[42] In part, this was because probationary periods were reduced, meaning workers become more expensive faster. Overall, one offshorer told the *Financial Times* that their production costs were eight per cent higher as a result of the new law.[43]

Despite the tightening grip of regulation, corruption is still a problem in many offshoring regions and this can stem from simply dragging down efficiency to entirely dominating the operation. One company found that it was forced into choosing the supplier for its components that the local government had recommended to them. It turned out that this was because the state had invested in that company and therefore was concerned with generating business for it.[44] At the more significant end of corruption are numerous cases of flagrant disregard for intellectual property in offshoring locations. In a notable outburst, Sir Anthony Bamford, Chairman of JCB, used the occasion of a trade delegation to Beijing to call China out on its willingness to turn a blind eye to patent infringement. He said:

> The machine designs of JCB products have been copied on many occasions by unscrupulous Far Eastern competitors. The practice is now commonplace in many industries. It is just not acceptable that one company's R&D effort is ruthlessly exploited by another elsewhere in the world.[45]

Indeed, the situation has been increasingly dire for multinationals. A 2010 survey of companies experiencing intellectual property issues found that 98 per cent of respondents with Chinese holdings suffered some form of fraud, with 26 per cent reporting IP theft or counterfeiting and 16 per cent reporting information theft. This was a rise from 86 per cent the previous year. There are many anecdotes about the levels of counterfeiting. One manufacturer of shampoo outsourced the bottle making to China and found that it was supplying the same bottles to another shampoo producer to make a knock-off version.[46] This form of infringement is most concerning for those offshorers attempting to break into local markets, where the counterfeiter often undercuts them and can ruin the brand through manufacturing cheaper, low-quality products. Companies with the whole production process offshored can see the problem spread beyond local markets. This allows counterfeiters to get hold of the finished products, original packaging and manufacturing paperwork that collectively enables them to sell the fake goods worldwide and in huge volumes. Disturbingly, this can make detection of fraud very difficult, even for high-tech goods. In May 2012, an investigation by the US congress announced that the US Air Force had bootleg products in its aircraft and that the main suppliers of Lockheed Martin, Boeing and Sikorsky were all using these unknowingly. Counterfeited night vision, GPS navigation and radio components were all discovered and 70 per cent of these were traced back to China.[47]

For those companies who offshored their R&D as well, the situation can be even worse. While it is simple enough to stop production of a product and remove the machinery from a plant, this cannot be done with knowledge production. For UK multinationals relying on their intellectual property for their competitive edge, this creates a convincing reason to onshore production for non-Asian markets, regardless of other benefits of offshoring, which are minimal from high-cost skilled labour anyway. According to the findings of EEF, the ability to innovate is the most important operation for these large companies, although it is important to note that this is about more than just R&D. Compared to China, where this activity is at risk, UK protection of intellectual property rights is perceived as 'good' by over 60 per cent of businesses, especially larger ones.[48] According to the Economist Intelligence Unit, there are strong links between the desirability of setting up R&D in a country and having thoroughly protective IP regulation, so this works in the UK's favour.[49]

Nexen

In the autumn of 2011, Nexen, a designer and producer of forklift trucks, announced its decisions to relocate its entire R&D programme from its manufacturing facility in Taiwan to the UK and to move the production of its popular 'X' range to a new manufacturing facility in Lowestoft, Suffolk. When it was founded in 2003, Nexen Lift Trucks established a joint venture with an Asian manufacturer to produce its forklifts under licence. Later that year the company decided to set up its own production facility in order to improve its competitiveness in the international market; it purchased the second largest lift truck manufacturer in Taiwan, creating the Nexen Motor Corporation.

Speaking on the 2011 move, Tim Mason, Nexen's UK Managing Director, stated that it would 'meet the concerns of several major component suppliers [who had been] reluctant to release specialised development items to Taiwan given the proximity of China'. Of course, Nexen's repatriation of its R&D and a substantial chunk of its production has not been motivated by a desire to protect its IP alone. Mason also cites the shorter lead times and higher productivity levels offered by manufacturing in the UK. Director Pam Oakes points to increasing labour costs in Asia coupled with the difficulties of recruiting a skilled workforce there. Nevertheless, Mason's thinly veiled reference to the issue of patent infringement highlights commonplace fears about unscrupulous Chinese competitors producing copycat designs, as well as the high value which UK manufacturers place on their intellectual property.

While corruption is being tackled in some areas, getting IP infringement cases to court and successfully prosecuting is still a very hard process. In part, this is simply because the required laws they would have at home are not in place in China, but there are also incidences of active cover up by government official with whom the counterfeiters have connections.[50] As such, some companies go to costly but necessary lengths to avoid enabling counterfeiting by dividing the manufacturing process between various plants or suppliers so no one can find out the whole process. Of course, the more suppliers and locations involved, the higher the cost and complication of production.

Logistics and other costs

Unless supplying the Chinese market, there is never an advantage in terms of geographical distance to the UK from offshoring to China. This might sound obvious, but it is also assumed that it can be overcome fairly easily. However, EEF analysed manufacturers' desire for proximity to customers and found that even multinational companies, despite their far reach, increasingly prefer to produce their goods within the same geographical location as the market they are selling to. Britain is, therefore, a key base for industry aimed at selling in the wider market including the rest of Europe, the Middle East and, often, Russia, and offshoring weakens this link. The incentive for this relatively localised production is twofold. Firstly, the costs of transportation are reduced, which is increasingly an issue as will be explained below. Secondly, the reduction in supply chain size reduces lead times, often to about two or three weeks instead of the six to eight weeks it takes to ship goods over from China. This allows local supply needs to be catered for, something which is becoming increasingly important as more and more companies only keep a few days' or even hours' worth of production materials. It also benefits the supplying company as well as preventing working capital being tied up in long journeys.

The length and time to deliver goods from offshore locations to the UK has been bearable for much of the 2000s, as lower labour costs outweighed this issue. However, the sharp increase in fuel costs is making this an increasingly painful exercise. Given that the majority of goods are transported from China to the UK via cargo ships, freighting costs are now fast eroding the savings made through outsourcing. In January 2000, the nominal spot price for European Brent crude oil was $25.51 a barrel and the cost of shipping tripled by 2008 when an oil spike occurred. This saw crude prices at a high of $132.72 a barrel and prices look set to reach similar levels again, leading to similarly expensive shipping costs: in February 2012, prices hovered around $120 a barrel.[51] Even without volatility (a nice thought), Boston Consulting Group still expects transportation costs to rise by 2.5 per cent per annum.[52] This is a double whammy for those companies that manufacture the base components in the UK and other locations and then send them to China for final assembly before shipping them back, incurring the higher costs on both journeys. For some firms, this in itself is enough to force production to be brought home. One such firm, Axminster Tool Centre, a family business based in the eponymous town, sells power tools but sells only £1 million's worth of British made tools while Chinese produced ones accounted for £27 million of sales. The Daily Telegraph recounted the experience of the company's managing director, Ian Styles: 'they had run the numbers on one tool racking system for workshops that they sell and found that for a £50,000 order they were paying £25,000 in freight costs. Add in the cost of holding stock and it makes domestic production more attractive. "I'd like to bring 10 to 20pc of what we sell back to the UK," he said.'[53]

For companies importing components, there will be additional costs from the need to store the incoming goods before they are transported to the UK factory. This is a cost often overlooked by companies. A case-study from the Institute for Manufacturing had to pay £195,000 to a hub that could look after this incoming stock. This cost though was 59 per cent higher than the projected cost the company had budgeted for, as a result of the longer lead times and compensating higher storage volumes the company now had to deal with.[54]

The time lag of around six weeks from order and production in China through to delivery in the UK can be a real penalty for a company that has to act swiftly to demand. Even for those that do not, if customers are prepared to pay slightly more for quicker delivery from rivals closer to them, this can also put the offshorers at a significant disadvantage. This tempers the enthusiasm of companies looking to move production further inland in China to avoid increasing labour costs, and the extra logistical difficulties shipping from central Chinese locations often presents further problems. The sleek infrastructure of the coast simply does not exist in these areas and according to the *Economist*, 'It can cost more to ship goods from the Chinese interior to the coast than from Shanghai to New York.'[55] Often, the most economic method is to ferry goods to seaports down the river, but this can add over a week to the delivery time. Moreover, for those companies that require other Chinese components, they often discover that the rest of their supply chain is based on the seaboard and moving away would present significant disruptions. Trying to bring a few senior managers and skilled employees would present further costs, as their upheaval would have to be incentivised. Many offshorers therefore simply do not have the 'luxury' of tapping into these cheaper Chinese regions.

N Brown Group

In late 2011, home shopping group N Brown, whose brands include Simply Be, Jacamo and JD Williams, hosted a meeting of potential UK suppliers. At the moment only one per cent of the group's products are made in the UK and their goal was to increase their UK production to at least five per cent. According to group development director Paul Kendrick, the major advantages of increased UK production will be quicker turnaround times - four to six weeks compared to twelve weeks in the Far East - and the possibility of ordering in smaller quantities: 'If a product made overseas does not sell well, then you're left with thousands of unsold garments but, by being able to order in smaller quantities here in the UK, there isn't that worry.'

In China, it has proven difficult for manufacturers to re-order popular products at short notice as factories are only willing to take on large orders. N Brown's initiative will enable the company to re-order quickly allowing them to commit to less stock initially. They will be able to test more products online, reordering when particular lines become popular. Domestic production will enable them to better respond to and profit from rapidly changing trends in fashion. Ben Lewis, CEO of high street retailer River Island, which increased its UK production by 50 per cent last year, has reported similar advantages:

> It has allowed us to get new fashion to our customers much quicker than we were able to, and as a result some of those products have become absolute bestsellers. We can get more of them and work closely with the factories. With clever design you can hold the price to something affordable.

However, onshoring of textiles is still being held back according to Kendrick: 'If I wanted 3,000 tops tomorrow, there's no-one to make them due to a shortage of workers in the sector'. Although N Brown has identified ten potential suppliers to supply basic jersey and knitwear garments, their more complicated products will still be made overseas.

While not necessarily a fault of the Chinese supplier, there are also issues with obsolescence that some firms face. The long transit time means that, while goods being shipped over are in perfectly good, working condition, they might not be wanted by the UK firm or its eventual customers. Indeed, Platts and Song found that in the tape measure manufacturer's case-study, £11,000 worth of their tape measures suffered obsolescence, as tape's yellow colouring was not right for the end market. They quoted the firm's managing director: 'If it had been made in England, we would not have made tens of thousands before we decided we didn't like them.'[56] Not only does it take longer from problems to come to light, but by the time they are, plenty of the goods could be in transit, which then have to be disposed of or stored at the UK firm's expense. Even then, the new goods will still take time to arrive, which could completely upset the supply chain. After exactly such a problem, another Platts and Song case-study had to purchase emergency components from its previous UK supplier, but at a much higher cost than usual given the small quantities. They ended up paying £18,000 for the British replacements of their Chinese goods that had cost £8,000.[57] This was a necessary investment given the inability to otherwise receive replacements on time to keep production going.

Like transport and labour, other costs like energy and raw materials incurred by companies that were once cheap are also becoming increasingly expensive. Energy, still cheap by comparison to the UK, is rising in cost and unreliability. Power shortages are ever more frequent as demand for electricity outstrips supply and in the first five months of 2011 alone, 20 regions of China suffered blackouts.[58] This has been caused by two principal factors. The rising price of coal has meant many power stations are no longer profitable, and power companies are reluctant to run their stations at a loss. Secondly, demand for electricity has risen as a result of more energy-intensive factories coming online. In 2010, Chinese electricity use grew by 14.6 per cent on the previous year, while installed power generation expanded by only 10 per cent and investment in the industry fell by 8.5 per cent.[59] This is clearly unsustainable, and without rapid investment, the Chinese grid will collapse in on itself. In 2011, there was a national gap of 18 million kilowatts, which the State Grid Corp of China described as possibly 'the worst electricity supply shortfall in history'.[60] This has the most notable effect on industry, which consumes almost three-quarters of China's power supply. As a result, the cost of electricity for industrial users was raised by 0.25 cents per kilowatt-hour, an increase of 15 per cent on 2010 prices. The higher expense is another addition to offshored companies' bottom lines, and the vulnerability of power supply is obviously a risk that has to be taken into account: losing power means no production.

Industrial land prices have also been growing rapidly in China, discouraging larger offshorers from buying land and constructing their own factories on the site rather than just outsourcing production. In part, this is a natural consequence of the overwhelming desire of companies to base their operations in the limited coastal cities with advanced maritime trade, which has pushed prices up considerably. Industrial land on average costs £68.20 per square meter, but this rises dramatically on the eastern seaboard, to £75.50 in the cheap city of Ningbo, £116 in Shanghai and £140.80 in Shenzhen.[61] The only way to find land below the average cost, or to avoid the high charges on the coast is to move inland, but as already discussed, this makes logistics much harder and further transportation costs will be incurred.

Moreover, the price of attractive land in China is significantly more than the price of the same unit of land in the manufacturing regions of the UK. A government survey of land prices in 2011 found that of the major cities, Newcastle is the cheapest by far, with industrial land costing £23.50 per square meter. Other industrial regions are similarly cheap, with the cities of Leeds and Sheffield costing £49.50 and £60 per square meter respectively.[62] For those who are just offshore outsourcing, or have not yet bought Chinese land, this might act as a significant incentive to expand production in the UK, especially when combined with the comparatively lower wages on offer there.

Similarly, the price of raw materials has increased dramatically over the past few years. As late as December 2007, a metric ton of iron ore was $36.63, but by March 2012 it had risen to $144.73.[63] More widely, an aggregate of industrial inputs price index (2005 = 100) has seen the cost of materials more than double from 71 points in March 2002 to 178 points in March 2012.[64] These have a very large eroding effect on emerging economy attraction given that everyone has to pay these costs, regardless of location. Given the lessening of price differentials, the value of quality production becomes more distinctive. Even as early as 2008, the Cast Metals Federation said that its UK members were receiving orders from customers who previously bought metal from Chinese foundries as a result of this.[65]

Currency issues

In China, the strength of the national currency, the yuan (RMB), has been growing in recent years in a trend which looks set to continue. In April 2012, one pound could buy 10.1 RMB but at the end of 2007, a pound could purchase 15 RMB and for the four previous years it hovered around this mark.[66] This sudden shift was the result of the Chinese government's caving in to American political pressure in 2005 and suspending the pegging of the yuan to the dollar, which then slowly floated upwards. While in operation, offshorers had obviously benefited from the peg keeping the yuan weak, but also from the confidence that it would remain so.

Now, however the UK firms pay for their Chinese goods, which depends on their individual contracts, they lose out from the strengthening yuan. If the payment is in sterling, then it is worth around a third less than it was around five years ago and Chinese suppliers have been raising their prices to compensate, although some lucky firms have escaped this. If paid in RMB, then the British firm has to pay a rising price for the same goods, which erodes its profits. Businesses will be even more worried by the fact that the yuan's volatility is still quite constrained by the government: this grip could lessen even further in the coming years. Even if this does not occur, it is still expected that the currency will appreciate by 3.5 per cent per annum through to 2015.[67]

In addition to the yuan's strength, the pound is also weakening anyway. A third of British manufacturers trade in sterling when conducting international trade and a fifth have no formal exchange management policy, leaving them highly vulnerable to rising import costs.[68] There is an additional complication for those firms that outsource production and pay for this in dollars, as some offshorers have to. The pound weakened relative to the dollar quite suddenly in 2008 from around $2 for £1 to around $1.45 for £1. This has not particularly improved over the last three years, with a weak $1.6 per £1 in April 2012.[69] On the plus side, this makes goods made in Britain much more competitive in the global markets and will increase exports, effectively rewarding those who have not offshored and incentivising others to return: given the increasing expense of importing, there will be more demand to invest in import substitution where possible. In addition, the UK becomes a more attractive location for foreign investors.

Similar to the exchange rate pegging, there was international resentment of China's export VAT policy that existed until June 2007. This was designed to encourage exporters to outsource or offshore to China, and meant that if their goods were not for consumption in the Chinese market (so shipped straight away), then they could claim a rebate on VAT, which was, on average, 13 per cent. In total, from 2002 to 2009, the Chinese government effectively subsidised exporting industries by a huge $455 billion.[70] Unsurprisingly, this was judged an unfair advantage by many countries and an infringement of WTO regulations. After increasing pressure and to avoid sanctions, the Chinese government altered the rebate rates of 2,831 goods in 2007, and while most were not fully eliminated, they were reduced to five per cent. Effectively, this meant Chinese companies were suddenly having to absorb the eight per cent difference or pass the cost along to British outsourcing customers, and British offshorers experienced the rise directly.

Hayter

In 2009 Hayter, the British lawnmower manufacturer, began repatriating production that it had previously outsourced to China. The main motivation for this was the weakness of the pound against the dollar. In 2000, Hayter's parent company, the industrial conglomerate Tomkins, sold the firm to a Chinese investment company and Hayter's new owner outsourced some of its production and assembly to firms in China as a cost cutting measure. At the beginning of 2009, approximately one third of the company's lawnmowers were being made in China and shipped to the UK.

However, the increasing Chinese labour costs of recent years and currency swings have made outsourcing less attractive for Hayter. The firm earns in pounds sterling when it sells its lawnmowers in the UK, but the costs of its outsourced operation are priced in US dollars. The pound, previously so strong against the dollar, has slumped, forcing up Hayter's costs in China. Further, a combination of rising wages and the appreciation of the value of the yuan reduced the formerly colossal labour-cost gap between the Chinese coastal provinces and the UK. Once shipping, inventory and other 'hidden' or risk-related costs associated with global supply chains are taken into account, the cost advantages of producing in China became marginal.

> In 2009 Hayter faced the prospect of production cuts following decreased sales in the wake of the financial crisis. As sales and marketing director David Sturges saw it, because the costs of operating Hayter's UK plant were fixed, it made most sense to use the plant at full capacity and to cut back outsourced production. This move also allowed the company to safeguard 175 jobs at the Hertfordshire plant.

National issues

At present, when British and Western firms in general deal with Chinese businesses it is their Asian partners who have on the whole been the junior role partners in any deals, which prevent them from advancing their production up the value chain. This is resented by many companies in China and the country as a whole will not be content to put up with being the Western world's basic goods supplier forever. Increasing numbers are looking to move up the supply chain and produce goods for other Chinese companies and consumers, potentially leaving new offshorers without suppliers.

According to research by CRESC, there are two principal ways Western companies dominate their Chinese partners.[71] They can allow Chinese ones to produce their goods while retaining the intellectual property rights and actual know-how of their competitive edge. These deals are often aimed at supplying goods in the Asian markets and can be seen especially in the automotive sector where the products of Sino-foreign deals currently supply 75 per cent of the Chinese domestic market but without room for exports. This way, the Chinese counterpart is unable to compete in the wider global markets and cannot develop or invest in this ability. Secondly, as the example of Foxconn showed (see p.46), the Chinese partners have their profits squeezed while the Western company makes a killing. They are hired to create products for the export market but derive comparatively little value from this: enough to survive and turn a profit, but not enough to invest in new ventures of their own away from assembling other companies' goods.

In part, this is the fault of the Chinese government, which recognised that China's economic boom was reliant on satisfying the needs of Western manufacturers and becoming the number one offshoring location. Its goal was simply to increase exports, without regard to the domestic market, whose demand it restricted, out of concern that this would cause labour costs to rise and therefore undermine foreign investment and provoke inflation.

Universities in China are also being used to increasing effect as part of the innovation process although at a general level of proficiency which remains below Western ones.[72] Now, with the increasing minimum wage, working directives and other developments necessary to create a modern, well-structured economy, China is turning towards a new economic strategy based on satisfying domestic demand and creating its own export market. Growth of the former can be seen in the semiconductor market. In 2005, Chinese demand made up a quarter of the global semiconductor market in 2005, and this rose to 32 per cent by 2008. Production of these products though rose faster, with China producing 10 per cent of its demand for microchips in 2004 and 22 per cent by 2008.[73] This is impressive growth, but if domestic demand begins to outstrip supply, offshorers might find themselves pushed out of the market.

Additionally, Chinese firms are beginning to create their own high-tech, high-value export markets that would compete with rather than work through offshorers. Many of these companies have utilised the knowledge and experienced they have gained by manufacturing Western companies' goods to speed up their development. For example, Vimicro Corporation, a Chinese manufacturer of microchips for digital imaging processes, designs chips used in half of the world's PCs and does so under its own name, not a Western company's.[74] It then outsources the production of these chips to Chinese production plants. This will squeeze UK manufacturers, not just in finding plants to outsource production to or import from, but also in terms of increased direct competition in new markets.

Lost in translation

A University of Cambridge study found that as a general rule, the further the distance between the offshorer and the actual location of production, the more unforeseen problems and risks the company is likely face and the harder it is to fix them.[75] It is undeniable that the language difference presents a significant barrier to effective communication. British suppliers talking to their Chinese suppliers frequently run into issues stemming from misunderstanding and just over 40 per cent of offshoring companies in a survey reported they had communication problems with their foreign plants, a higher number than even those experiencing quality issues.[76] While confusion is not necessarily as damaging as production problems, this is complicated by the cultural trope of saying 'yes' to questions to show one is listening, and not as confirmation of understanding or acceptance and this could lead to dramatic problems down the line.[77] Often, the lack of comprehension means British firms have to be unduly careful, rely on email and ask questions one at a time in the clearest language possible. At the same time, Chinese firms can be reluctant to communicate failure to their customers and do not let them know about delays, quality issues or even problems with the production process itself.

Distance, both physical and cultural, certainly does create obstacles and according to some British businesses, extra ones for those looking to outsource rather than offshore. To find a viable Chinese firm to outsource work to, there is a lengthy process of building up a relationship before anything can actually happen. Frequently, the nurturing of the relationship requires meeting in person, which requires considerable travelling. While it is possible to find firms over the internet that would happily begin production straight away, these are the ones where delivery/quality is more questionable and they might not have the skills and equipment that they claim. It is certainly worth investing time and money in making sure the right supplier is chosen, but this is less convenient given the distance and thorough analysis is often only taken by the most cautious companies. For larger firms, this is less of an issue, especially at the beginning of a potential long-term business partnership. However, for smaller firms this could prove an expensive and time-intensive process. Platts and Song found that 'the cost of time for search for, visiting and negotiating with suppliers represented a significant, but widely variable, percentage of the set up cost' with a modal range of 45 to 58 per cent.[78]

Companies in the Far East work on different pay time-scales to those in the UK. While UK companies often set payment terms as between thirty to sixty days, in Asia this could be as short as five to fifteen days, and there are anecdotal claims that some companies will not even start production until being paid in full.[79] This creates disruption to usual payment cycles of the offshoring firm and creates additional risk: if the product is of dubious quality, then the payment might have to be made before the goods are even received, making it harder to negotiate over compensation or supplying a new batch.

Where firms have offshored production but retained UK R&D departments, there are additional difficulties. As well as being geographically distant, the researchers are isolated in mental and social terms from the products they design, since they require other people to make them a reality. For example, while a new form of technology could appear to work well on paper, the workforce producing it could find it would be unviable for them to create and that it is in need of adjustments, revealing their less academic but equally useful form of knowledge. This is less possible though in an environment where the production workforce is unable or unwilling to point these issues out. The first the British company could hear about the problem would be when the shipments arrive and the products fail to perform. Retaining production in the UK, even if still physically separated from R&D facilities can prevent this from happening and ensure any product that leaves the drawing board does so without problems.

Turning to other low-cost countries

At first glance, it might appear easier for companies to simply uproot from the increasingly expensive developing nations and transfer production to other ones that continue to offer cheap labour. This is not as viable as it appears. The advanced infrastructure available in places such as the east coast of China is simply non-existent in other countries like Vietnam, Bangladesh and Cambodia. Vietnam for instance generated $11.6 billion in foreign direct investment through 2011 and an impressive GDP growth level of 5.9 per cent.[80] However, the rapid growth has also led to double-digit inflation, which has yet to be curbed, and infrastructure continues to be underdeveloped. The same is true of most other low-cost countries and whatever the labour costs, if it is too difficult to actually ship goods and construct factories, there will be no demand to set up shop there. The investment that has occurred is from large multinationals producing high-volume labour-intensive goods with enough clout to overcome these directly or indirectly through leaning on the state for improvements. The average British SME will not be catered for.

An additional problem with other low-cost countries is the size of the workforce. China and India both offer the huge workforces required for the sort of large-scale mass production that can overcome labour productivity issues: Foxconn employs 1.2 million workers on the Chinese mainland, making it the tenth largest employer in the world and the largest manufacturing employer by far.[81] Other countries simply cannot supply a workforce on this scale: Indonesia is the third largest Asian country in terms of population with 237 million citizens, but pales in comparison to China's 1.3 billion and India's 1.2 billion inhabitants. The only companies that will be able to benefit from moving production to other Asian countries will be relatively small-scale, without a need for large workforces and able to cope with underdeveloped logistics.

2

The United States as a case study

Onshoring and encouraging onshoring in America

The US offers the UK a glimpse at the effects and benefits onshoring industry can have. It is an ideal case study because, while America is obviously much bigger than Britain, its economy works roughly along the same lines and its manufacturing sector occupies very similar shares of its economy. In 2010, manufacturing contributed 12 per cent of GDP (same as the UK) and manufactured goods represented 61 per cent of all US exports (52 per cent in the UK).[1] America has experienced the onshoring trend for slightly longer and certainly places more importance on bringing industries home. The economic tools it has pioneered, and the difficulties it has run into while trying to encourage repatriation all act as guidelines for Britain and its government.

Like the UK, the US offshored a sizeable proportion of industry, with American manufacturing multinationals employing two-fifths of their workforce outside the States (compared to an average of a third in other sectors).[2] Many smaller businesses also offshored and outsourced production to take advantage of emerging economies and while this was the case for many years, the trend accelerated as time went by. Already though, America has begun to experience the repatriation of some of these businesses, not least through government encouragement. Manufacturing is still an integral part of the American economy. In 2009, the US was estimated by the World Bank to have a 20 per cent share of worldwide manufacturing, placing it slightly higher than China.[3] Since the recession, the US has seen its manufacturing employment grow faster than any other developed economy and in 2010, it gained 300,000 industrial jobs.

The health of American manufacturing is primarily due to the country becoming much more attractive and competitive as a place to set up factories as a result of sustained productivity growth while wages and the strength of the dollar have fallen. Nationwide, US wages grew by 4 per cent in 2005-10, compared to 19 per cent in China over the same period.[4] The price of energy has also fallen due to the shale gas market opening up, with gas prices under a quarter of UK equivalents. Onshoring has also been aided by the increases in shipping costs. The American Institute for Economic Research examined the case of a manufacturer of toilet paper dispensers called Xten Industries Inc. Now producing their products in Wisconsin, it found:

> A standard 40-foot shipping container typically holds 2,000 of the type of dispenser Xten makes. At the current shipping price, it would cost $3 each for a Chinese-produced version of the product to reach the US market, where its retail price is less than $15. Even if labor was free in China, it wouldn't make sense to produce the dispenser there.[5]

On the other hand, the 'push' factors from China have been increasingly apparent. Rising costs in China mean that Chinese prices relative to US levels in PPP terms were 50 per cent in 2009, a rapid rise from under 36 per cent in 2002 and a sharp increase even since 2007 when it was 42 per cent, due to the combined forces of the recession in America and rising offshoring costs.[6] By 2015, the Boston Consulting Group has estimated that for goods produced for the North American market, it will be as economical to make them in the US as it will be in China.[7] Bank of America Merrill Lynch has also predicted that the chemicals, food & beverage and computer & electronics industries will repatriate production as: 'these growing industries possess characteristics that make offshore alternatives less attractive or unviable'.[8] With multiple industries at a tipping point where onshoring is becoming possible, government incentives can tip this in America's favour.

To show how important this is for the American economy, and how central it is to his Administration, President Barack Obama made onshoring a central tenet of his 2012 State of the Union address:

> ...Long before the recession, jobs and manufacturing began leaving our shores. Technology made businesses more efficient, but also made some jobs obsolete...
>
> ...No, we will not go back to an economy weakened by outsourcing, bad debt, and phoney financial profits. Tonight, I want to speak about how we move forward, and lay out a blueprint for an economy that's built to last – an economy built on American manufacturing, American energy, skills for American workers, and a renewal of American values...
>
> ...We can't bring back every job that's left our shores. But right now, it's getting more expensive to do business in places like China...

> ...So we have a huge opportunity, at this moment, to bring manufacturing back. But we have to seize it. Tonight, my message to business leaders is simple: Ask yourselves what you can do to bring jobs back to your country, and your country will do everything we can to help you succeed.[9]

These extracts, and the wider desire to increase onshoring, formed the principal economic ideas in the speech, which in itself is the President's opportunity to discuss their agenda and national priorities for the year. In other words, onshoring is being taken very seriously by the Obama Administration and is seen as central to economic revival. The example Obama used in the speech, Master Lock, a producer of combination locks in Wisconsin, repatriated its production from China, and it found that the six-fold productivity increase this gave the company actually meant total production costs fell. There has been nowhere near the same level of discussion in the UK, and political speeches have barely even referred to onshoring. The US is ahead of Britain in recognising the trend and its actions can therefore be seen as a good model for us to follow.

The Obama and wider government enthusiasm for onshoring is clearly not just rhetoric and the policies are also backed up by opinions from the business world. A survey of 3,000 American manufacturing SME executives found that 85 per cent 'see the possibility of certain manufacturing operations returning to the US'.[10] The survey, by Cook Associates also gave a breakdown of the factors with the most influence over any onshoring decision:

> ...37 percent cit[ed] overseas costs as the major factor... nineteen per cent cited logistics and 36 per cent stipulated other reasons, including economic/political issues, quality and safety concerns, patriotism and overseas skills shortages for highly technical manufacturing processes.[11]

These results are particularly intriguing given that two-thirds of the companies polled currently outsource some or all of their manufacturing. Some of them are clearly intent on bring this home, many for the reasons already discussed above.

Crucial to this process is active incentivisation and America has also been better than the UK at creating rewarding packages for businesses to relocate back to the US. In June 2011, Obama launched a $500 million package specifically to generate more employment in manufacturing through developing innovation and entrenching competitive advantage. Interestingly, this package was also aimed as much at low-tech industries, as high-tech ones: all that matters is their high-value added economic contribution. Tax incentives were also created in January 2012 that included:

- Scrapping tax breaks for those who offshore manufacturing.
- Making companies pay a minimum tax for profits and jobs overseas.
- Giving a 20 per cent income tax credit to allow for the expenses of shifting operations back to the US.
- Doubling the current 9 per cent deduction on domestic production activities to 18 per cent for advanced manufacturing.

- Making permanent an expanded tax credit for R&D conducted within the US.

There are also incentives given to individual companies to help them onshore production. US Electrolux was given financial support by both the state and local governments worth $137 million to set up a new plant in Memphis, Tennessee to make cooking appliances. Other additional benefits have potentially raised the total support to $188 million, which of a $190 million cost for the new factory is clearly a huge incentive to set up long-term production there.[12] The federal government itself is getting actively involved in courting businesses. Nissan was given a $1.45 billion loan by the US Department of Energy, through its Advanced Technology Vehicles Manufacturing Program. This covered most of the $1.8 billion bill to build a new plant in Smyrna, Tennessee. As these examples suggest, it is not just SMEs that are returning production to America, and this is very important. While Apple might be the most famous large company that refuses to onshore its production, this is by no means the attitude of all large businesses.

The US government has also funded an organisation called the 'Reshoring Initiative', which aims 'to bring good, well-paying manufacturing jobs back to the United States by assisting companies to more accurately assess their total cost of offshoring, and shift collective thinking from "offshoring is cheaper" to "local reduces the total cost of ownership."'[13] The site has launched an online tool dubbed the 'Total Cost of Ownership Estimator', which converts a producer's costs and risk factors into a single figure to encourage them to make more objective sourcing decisions. As has been seen in the UK and the US, many manufacturers have failed to make their offshoring decisions based on a holistic understanding of the total costs of production (that is, one which allows for the hidden costs of offshoring). The Initiative is based on the idea that manufacturers who calculate costs completely are far more likely to outsource to local domestic firms. One such company who benefitted from this was the Illinois-based Morey Corporation, a manufacturer of circuit board components. The company reduced its inventory costs by 94 per cent after onshoring production back from China while also improving quality.[14]

Even without government support (and therefore based on the general attractiveness of the American economy), Mars Chocolate North America made headlines last year when it announced it would be building a $250 million chocolate factory in Topeka, Kansas. Creating 200 jobs, this is the first such domestic plant Mars has opened in 35 years.[15] Indeed, other states similar to Kansas, such as Arkansas, Mississippi and Nebraska have seen the bulk of onshoring, as manufacturing wages are around ten to fifteen per cent lower there compared to the American average.[16] In addition to increasingly flexible unions willing to negotiate over future pay, this has reaped large benefits. In October 2011, Ford announced that it would create an additional 12,000 hourly jobs and $16 billion of investments after reaching an agreement with United Auto Workers, a crucial trade union. Part of this will involve insourcing jobs from Mexico and China.[17] Mexico, however, will continue to receive a portion of the business leaving China. The Boston Consulting Group estimates that by 2015, Chinese workers will be earning

25 per cent more on average than their Mexican counterparts and as a signatory of the North American Free Trade Agreement, companies in Mexico will also be able to export their products to the US without paying duty taxes.[18] High-value work is still likely to be repatriated to America though, if it leaves China.

The companies that began onshore production early appear to have done well out of the then-unforeseen consequences of the recession. One such firm, Peerless AV in Aurora, Illinois, repatriated production after it saw counterfeited copies of its patented metal brackets and stands for televisions on sale. The concern around intellectual property led Peerless to return production home as the recession hit, and this allowed it to buy discounted equipment and machinery, as well as a workforce with the skills it required. The proximity of suppliers and customers meant the company could now produce a new product in a couple of weeks, as opposed to six months when production was still China-based.[19]

In addition to the onshoring that has already happened, Boston Consulting Group (BCG) has examined the future trends of seven primary industries that account for $200 billion's worth of imports from China alone to the US and predicted their path over the next decade. Labelled 'tipping point' sectors, because their rising production costs in China would mean production in the US for the US market is cheaper in five years, the seven industries were computers and electronics, appliances and electrical equipment, machinery, furniture, fabricated metals, plastics and rubber, and transportation goods. Within the seven industries, a shift back to US production has already begun, but with a change in company style. There are increased numbers of lower entry-level wage rates that compete with China when productivity gaps are taken into account. To offset this, some companies have created more fluid hierarchies, allowing worker promotion to occur faster. In total, these represent around two-thirds of all US imports from China. BCG calculated that ten to thirty per cent of the goods these industries produce could return to the US in the next ten years, which would contribute an additional $20 to $55 billion to the American economy. Beyond the direct job gains within these industries, other roles will also be created within the wider sectors associated with the repatriation: in construction, logistics and retail to name a few. In total, BCG expects this will add two to three million jobs and reduce unemployment by 1.5 to 2 percentage points.[20] If a similar proportion of British offshored industry could be repatriated, to serve the British and European markets, this would be the equivalent of creating 300,000 to 450,000 additional jobs.[21]

Apple

Apple's heavy reliance on overseas production is world famous, as is Steve Jobs' retort to President Obama when he asked what it would take to have the iPhone manufactured in the US. The answer by Apple's ex-chief executive was simply 'those jobs aren't coming back'. Despite being one of the world's largest companies, Apple's refusal to negotiate over this at present is a thorn in the side of the US and gives good reason to somewhat temper enthusiasm for onshoring: it shows America cannot expect all businesses to onshore, even when the economic climate is in their favour. As one existing Apple executive has said: 'we don't have an obligation to solve America's problems. Our only obligation is making the best product possible'.[22] Nonetheless, Apple is a useful model to examine in assessing the potential for onshoring manufacturing, because it reveals how little value Chinese production adds to their goods and the minimal impact onshoring production would have on their already considerable profits. It also shows exactly where the advantage in offshoring lies for large companies, and the barriers that have to be overcome to tempt them to return home.

Apple employs around 43,000 workers in the US and a further 20,000 overseas. Given its huge value, this means each employee is earning the company roughly $400,000 in profit, which is more than Goldman Sachs or Exxon Mobile. However, this is only possible because of the 700,000 people employed by the companies which Apple contracts to actually produce its goods. The *New York Times* recounted the value of offshore production for Apple, when last minute changes to the iPhone's screen meant an assembly line overhaul by its Foxconn plant:

> New screens began arriving at the plant near midnight. A foreman immediately roused 8,000 workers inside the company's dormitories... Each employee was given a biscuit and a cup of tea, guided to a workstation and within half an hour started a 12-hour shift fitting glass screens into beveled frames. Within 96 hours, the plant was producing over 10,000 iPhones a day.[23]

While production costs in China are increasing, anecdotes such as this imply that large companies can still lever their workforces in ways that Western employees would likely not tolerate. A former Apple executive suggested that the speed and flexibility of these Chinese factories meant 'there's no American plant that can match that'.[24] This is where the value in offshoring continues to lie.

Nonetheless, what is good for Apple is not necessarily good for America. In 2009, importing the iPhone to the US from China alone created a trade deficit of $1.9 billion, even accounting for the American components included in the finished product. This was 0.8 per cent of the entire Sino-American trade deficit.[25] However, this huge sum hides how little China actually contributes to the production process. In 2009, the cost of materials for the 3G iPhone came to $172.46 and because the iPhone is just constructed in China, this means the valued-added Chinese labour adds to the product is worth $6.5 and constitutes just 3.6 per cent to the total manufacturing cost. The other 96.4 per cent comes from Germany, Japan, Korea, the US and other countries, where the components were made. Removing these from the equation means the actual trade deficit from Chinese assembly should be worth only $73 million. Given the large profit margin on the $500 iPhone, which was 64 per cent in 2009, Apple is able to produce a tidy sum for itself on each sale, which is simply not possible in a fiercely contested market and therefore discredits the idea that Apple relies on cheap Chinese labour for competitive reasons. The same is true of the latest 4G iPhone model. Components cost $171.35 with Chinese labour at a further $7.10. With the phone selling for as much as $630, Apple makes up to $452 on each unit sold, an even larger, gross margin of 72 per cent.[26] All this suggests Apple is driving down manufacturing costs via offshoring for its own benefit. This is reinforced by an investigation into another Apple product, the iPad. Selling at $499, with profits of $150 for Apple, the 16 gigabtye Wi-Fi enabled iPad contains components worth $154, from America, Japan, Korea and Europe, with labour costs of $33, but again China's contribution was paltry, at only $8.[27]

An investigation into the feasibility of producing the iPhone in the US was conducted by Yuqing Xing and Neal Detert of the Asian Development Bank Institute. They concluded:

> If all iPhones were assembled in the US, the US$1.9 billion trade deficit in iPhone trade with [China] would not exist. Moreover, 11.4 million units of iPhone sold in the non-US market in 2009 would add US$5.7 billion to US exports.[28]

Xing and Detert calculated the additional cost Apple would face if it had retained an American workforce to put together the iPhone in 2009. Assuming (in China's favour) that wages were ten times higher in the US and equal productivity between the two countries, assembly costs would be $65 and at the same selling price of $500. This still means a 50 per cent profit for Apple, which few would argue is a raw deal. Since these calculations, a further study has been undertaken by CRESC, which calculated that if the 4G iPhone was constructed in the US, Apple's profits would fall to 46.5 per cent, if assuming an average electronics industry wage of $21 per hour, which would still make it the most profitable phone in the world. The report summarised:

> Certainly, Apple should not be an object of praise and emulation because its business model is not generalizable

> without harm to the US and limited benefit to China. Financialization means many things but, inter alia, it denotes the absence of judgement which celebrates high financial returns at one point in a chain as a brilliant success.[29]

For a company sitting on a pile of money worth almost $100 billion, it is clear that self-interest is the only thing holding back US production. Despite Apple's refusal, the theoretical onshoring of their manufacturing would clearly be of huge benefit to America, especially since the assembly workers required would be low-skilled employees, a stratum continuing to struggle to find jobs. Given the new management of Apple, and the continuing pressure of corporate social responsibility that has led to investigations of Foxconn's practices, the onshoring of Apple might be a high-profile possibility that should be revisited in the near future, by both the company itself, and the US government.

Warnings

Despite being a model for the UK, and showing the British government the potential for onshoring and how to encourage this, the earlier start of the American onshoring trend means it already provides some warnings for the UK about the bottlenecks that could constrain repatriation. According to the President of the Federal Reserve Bank of Cleveland, the US manufacturers that weathered the recession well were often those involved in machinery and equipment production. This was because many manufacturers in America were looking for ways to maximise or protect their profits in the time of crisis. They reduced their bottom line through investing in new tools and improving productivity.[30]

While this was good for the companies, those re-entering industrial employment often found that even if they once worked in the same sector, the technology had moved on since business moved away and returned to the US. As such, many workers are now unfamiliar with new production processes and have to be trained up before they can use them. This disadvantage can be offset, in part because the basic skills of workers mean they pick up the necessary knowledge fast, but also because the required workforce is small: advanced machinery means production is less labour-intensive. Research by Wells Fargo found that: 'from 2004 to 2006, output in [US] manufacturing by insourcing manufacturing firms increased 19 per cent, while adding only 2 per cent to payrolls'. The main cost was unsurprisingly in capital expenditure, which rose by 10 per cent.[31]

However, for the American government, this creates an additional issue: the average onshoring manufacturer is looking to employ fewer, more educated workers than they once did. Figure 5 (p.185) shows the historical trend over time, with a loss of about 4 million manufacturing jobs over two decades. Rolls-Royce is an example of the pattern of employing fewer, higher skilled workers: it announced with much fanfare that it was creating '600 highly skilled jobs in Virginia' and even received a Presidential visit in March 2012 to the factory where these workers will be added.[32] However, it has shed many more in the way of lower skilled jobs, so the end result is net job losses.[33] As a whole, the Bureau of Labour Statistics estimates that from 2010 to 2020, manufacturing employment in all its guises will shrink, although only by a very small amount: from 11.52 million jobs to 11.45

million jobs, a percentage change of just -0.6 per cent.[34] Onshoring cannot be considered the panacea to our economic woes.

The advanced machinery usually operates via computers, so this also means companies need highly IT-literate employees. There is a demographic impact to this. While there is a large pool of older, highly traditionally-skilled workers, many companies feel they lack the ability or adaptability to work with new production processes. For example, Chesapeake Bay Candle used to offshore production of its scented candles to China and when the US put an import tariff on Chinese candles, then from Vietnam. However, Mei Xu, owner of Chesapeake Bay Candle and sometime advisor to the White House, then pulled production back to the US, in part because of rising labour costs, but also so the plant could be near her laboratory so that production could respond to new trends faster. This close research/production relationship means workers need more skills. Xu said: 'workers in their 50s may be willing to work, but may not have the computer skills to use the technology'. Xu's answer is to utilise the incoming entrants to the workforce, but this ability is constrained by their lack of experience. She said 'we really need high school kids to get some vocational training'.[35] This lack of real attraction, from either the younger or older workforces appears to be the greatest challenge to the US unlocking a larger wave of onshoring, especially as Chinese workers are increasingly *au fait* with high-tech production processes. Wells Fargo concluded a report into onshoring:

> A manufacturing worker today needs to possess a much higher skill-set than in the past and must continue to do so in the future. This higher skill requirement is the challenge for America, with many manufacturers reporting that they cannot find qualified workers due to these higher skill requirements. In order to maintain our status as a key manufacturing nation, a greater focus on education and skills upgrades will be required.[36]

The long-term employment trends appear to back this up: blue-collar and middle-skill jobs in manufacturing have decreased significantly: only college-educated worker numbers have risen and there is a large demand for qualified engineers.[37] Even Steve Jobs, having told Obama there was no way to repatriate Apple's production, conceded that it could bring some of the most skilled manufacturing home if America could train more engineers.[38] To oversee a 200,000 strong workforce of low-skill assembly-line workers, Apple estimated it would need 8,700 high-skill industrial engineers, but that at current rates, it would take nine months to find this quantity in the US. When searching in China, it claimed this took just 15 days. The problem it would appear is not one of quality, but of quantity. Large-scale production requires a decent-sized workforce.

Thankfully, it is likely that this is less of a problem for UK industries. For the most part, they are not as large as American counterparts and we have no company akin to Apple in terms of size. While this might not appear to be a beneficial situation, it does mean the UK is less likely to find the size of a required workforce a barrier to repatriating industry. Britain has only 205 manufacturing companies that employ over 1,000 people, which pales in comparison to the US's 2074 firms.[39] Obviously, the US is able to operate on much larger scales than Britain can dream of, with 283 American companies employing more than 10,000 people. At these huge sizes, Chinese production does retain its appeal for longer. Nonetheless, there is a real need for better vocational training in the UK to encourage more firms to return production home, but as shall be seen, these are likely to be in specialist areas.

3

What sorts of companies are most likely to onshore back to Britain?

Introduction

The rush to offshore was rapid and frequently herd-like: companies saw their rivals offshoring and assumed this was the secret to success. Hindsight shows this was not the case, and experience will make companies wary about rushing into decisions again: another fruitless relocation, even back to the UK, would bankrupt many companies. Instead, manufacturers will be looking at long-term trends, and where they will be setting up shop for at least the next decade or two. These are the sort of companies the UK wants to encourage to return production home, and how this is possible will be discussed in the next chapter. Here, we will concentrate on the sectors and types of company most likely to repatriate production, either through insourcing or outsourcing to British firms. These companies can be divided into two rough groups: the component manufacturers who would be bringing the production of their own intermediate goods back to the UK; and the companies producing finished goods, who would turn to UK suppliers rather than foreign ones.

The report into onshoring by Cook Associates found that 'low-volume, high-precision, high-mix operations, automated manufacturing and engineered products requiring technology improvements or innovation' were the most likely to return to the US.[1] This is unsurprising and is just as true for the UK as America. These are the manufacturing operations that do not benefit from the cheap but simple mass production abilities of developing countries as they need a high level of overseeing. The UK has the additional advantage of not having a direct next-door low cost neighbour to compete with for business. The US has to contend with Mexico, which offers cheaper labour and without the logistical costs and complexity of offshoring that a move to Central and Eastern Europe would entail for British companies. Nonetheless, these European locations are offshoring threats that will grow if Britain becomes complacent and loses its attractiveness. For now though, Britain already has a global reputation for producing high-quality products, and it is this competitive arena in which onshorers can thrive. In addition, they might work within niche markets, deliver flexible production and strengthen overall supply chains. Each of these is a facet of quality production, beyond the high quality of the product itself.

Of course, not all sectors are in the same situation, and the degree to which they will continue to rely on overseas production varies. Some, such as manufacturers of optical, electrical and mechanical equipment utilise foreign components much more, and could keep the final assembly of products in-house or onshore this, even if they cannot insource any of the actual parts. At a very basic level, any opportunity presented by onshoring that will reinforce an existing competitive advantage, or can create a new one, can be the start of the repatriation process. The push or pull factors relating to core competencies will have an overriding effect on the decision. The same is true for those outsourcing production, who could well find that British suppliers offer a greater capability, which in turn is effectively a new core competency.

It might sound obvious, but the first companies to onshore will be those that never left. Some UK companies did not put all their eggs in one basket and offshored some production while retaining domestic ability as well. During the present times of lower demand, firms that did this are finding it easier (and financially safer) to shut down the offshored plants and return British facilities to maximum output. Beyond this, other British companies currently wanting to increase production, that might once have offshored production to China, may now look at the situation and conclude that this is no longer a desirable option. Instead, they expand their UK facilities or decide to outsource to other UK firms. The slowing migration of companies overseas is evidence of this trend, as is the growth in low-tech but high-value British manufacturing. These low-tech companies account for very little in the way of direct exports and in 2008, high and medium-high tech goods accounted for two-thirds of all British exports. This belies the increasing domestic presence of UK-based basic component firms, whose products are used by the advanced manufacturers. This means that much of the value of the suppliers is actually hidden within the export revenue of their customers and the onshoring trend can be seen through this. From 1995 to 2008, low-tech companies' share of exports increased only from 31.5 per cent to 34.5 per cent, despite output increasing far more than this. This strongly implies that the increased output has been put to use in increasing in the domestic supply of intermediate goods. This is not particularly surprising given the different outputs of this manufacturing sector. Those such as food and sheet metal have little export demand, as most countries produce these goods to consume domestically. The fact that Britain still imports a large amount of these goods implies though that there is still a market for British production to fill and it is by no means saturated yet.

Those producing high-quality goods

The recession has led many UK companies to question their business models, and work out how best to survive in the global supply chain where competition is fierce. Many have decided to place renewed emphasis on the quality of their product or the service provided to the client as ways to distinguish themselves from the crowd. Worldwide, the label 'made in Britain' continues to evoke the idea that British manufacturing produces goods of high quality and finish: products worth paying a higher price for. Compared to roughly half of companies seeing quality as an issue in offshore production, only six per cent of respondents to EEF's latest survey felt British quality is bad, and 80 per cent saw it as good.[2] In most cases, the automatic association of the UK with quality acts as free branding for companies and is the envy of many other countries. Most businesses producing goods in the UK have capitalised on this and entrenched high quality as a significant selling point. This means that, by onshoring production from China, firms can send a message to customers that they are serious about maintaining quality and can then move their products upmarket or command a higher price to offset higher labour costs.

In some sectors, it is the quality of British goods and the services provided by manufacturers that drives demand for their products. For example, a British turbine blade coming from the Rolls-Royce factory in Derby has to be able to endure spinning at 12,000 rpm at 1,600°C, which is 200°C hotter than the metal's melting point and it has to have a lifetime of 15 million miles.[3] While still essentially a shaped piece of metal, it is obvious that the blade is so much more than this, and the creation of a structural composition of the blade that allows it to undergo such stress is the source of the blade's competitive advantage, which allows the overall jet engine to run more efficiently and reduce fuel consumption. The Government is often unaware that high-quality high-value does not necessarily mean a complicated and electronic product.

The blade is also a good example of the importance of the manufacturing process as well as the end product. It has a hollow centre which is accessible to air through thousands of tiny holes, and this is what stops the blade from melting. As no normal drill is precise enough to make these holes, Rolls has a created an advanced electronic process to create them. It is this that creates the very high value of a 'simple' product. This process could be carried out anywhere in the world, including China, because the cost of the machinery is essentially the same worldwide. However, the expense of the equipment, which would dwarf labour cost savings, and the entrenched position of Britain as one of the world's foremost advanced manufacturers keeps UK production more attractive.

Mulberry

The UK's luxury goods sector is experiencing significant growth. Mulberry ranks highly among the British retailers for whom strong demand is being fuelled by a growing, affluent Asian middle class with an insatiable appetite for branded Western goods. In September 2011, the company expanded its factory in Chilcompton, Somerset and in 2013 will open a second factory in nearby Bridgwater, providing 60 and 256 new jobs respectively. The second UK factory will produce approximately 140,000 handbags per annum, primarily for export markets, and is being financed by a £2.5 million contribution from the government's regional growth fund as well as £5 million of the company's own capital.

In the 1990s Mulberry moved 50 per cent of its manufacturing overseas to Spain, Turkey and China. Currently, only 20 to 30 per cent of its bags are made in Chilcompton. But with brands that emphasise British heritage (including competitor Burberry) weathering the recession well, it is perhaps unsurprising that a luxury brand is looking to repatriate some of its production to the UK.

Indeed, Mulberry's website proudly proclaims that its Somerset factory, The Rookery, 'is the only leather goods factory of its size left in the UK'. These brands have capitalised on the publicity and hype around the Queen's Diamond Jubilee and the London 2012 Olympic and Paralympic Games.

Mulberry is also a member of the Sustainable Luxury Working Group 'comprised of companies in the luxury industry that are committed to advancing good social, environmental, and animal-welfare practices in their business operations, including sustainable sourcing practices'. This is particularly manageable through domestic production which effectively guarantees ethical practices, another facet of quality

Expanding British production has required investment in training, which Mulberry has embraced, with a view to continued domestic output. In 2006, 50 per cent of the Chilcompton factory's workforce was over the age of 50 and 13 per cent were over retirement age. Now each year its apprenticeship scheme trains eight young people from the local community for a qualification in Leather Goods Manufacture to create: 'a new generation of talented artisans to follow in the footsteps of those craftsmen and women who have helped build Mulberry's reputation for luxury craftsmanship'.

Those working in niche markets

Britain has traditionally been good at competing in niche markets and this trend is continuing to develop. The targeting of small markets is the source of the high value of many British companies' products and, on the whole, UK firms are better than their rivals at focusing their skills and tailoring their innovative qualities to the customers' needs to generate constant demand. The EEF 2007 survey found that niche production as a means to a new competitive advantage has been a rising strategy among businesses, with 45 to 53 per cent of businesses adopting this from 2004 to 2007.[4] This is especially prevalent among British SMEs, over three fifths of which adopted a niche market as part of their strategy for competing, compared to two fifths of large companies.[5]

High-quality and niche production go hand-in-hand for British manufacturers. There are still many industries where Britain perhaps contrary to expectation, is still competing strongly. Chemicals and chemical products increased their export output from £23.9 billion in 2000 to £40.7 billion in 2007. Even the decline in older industries in real terms has been slower than it might have been. A key example of these would be the textiles, leather and footwear manufacturing industries where total exports over the last nine years grew, from £6.1 billion in 2000 to £6.6 billion in 2008, although this was a slight decline in real terms.[6] This slow reduction appears to fly in the face of the 'Primark effect', which would predict a rapid movement overseas of cheap mass produced goods. The success of the British textile industry is partly due to its having avoided trying to compete in this market, where on a cost basis it would never win. Instead, it has forged a niche for itself as a purveyor of fine quality goods and it is this quality of British textiles that sells the product and makes it worth exporting. For example, Savile Row is synonymous with the concept of superior British clothing and this justifies the price to the consumer. Moreover, this excellence has sold the industry overseas and has brought in an international clientele along with a self-sustaining trade that will continue as long as the quality does.

Mini Gears

The recent experience of Stockport-based gear manufacturers, Mini Gears, demonstrates the way in which a focus on niche markets can enable UK manufacturers to remain competitive in spite of low-cost overseas competitors. In the past, the company tended to manufacture medium to high-volume batches of commercial gears and racks. However, for a decade, low-cost Chinese competitors steadily eroded its customer base, to the extent that Mini Gears decided to establish an additional revenue stream by sourcing components from overseas.

However, in recent years the company has adopted a new strategy, using its specialist knowledge and skills to diversify into low-volume and high-precision industries including racing cars, gears for classic cars, high-end valves applications and rotor pumps. For example, Mini Gears has historically produced gear racks for portable drills, electric generators and stair lifts.

These parts are usually made in large volumes and are very price sensitive as a result. More recently the firm has found success by moving into niche markets, targeting low-volume, high-quality gear racks for areas including commercial aircraft seating. For customers ordering in lower volumes, high quality trumps low costs.

Furthermore, according to company chairman, Paul Darwent, the advent of escalating costs in formerly low labour cost countries has led to several new enquiries for parts that would usually be sourced overseas. As a result of the rising cost of producing in China, Mini Gears has won back a number of medium to high-volume contracts for its UK manufacturing facility since mid-2011.

Those producing rapidly changing products

Part of what high-quality companies expect from British production is the ability to rapidly change products to suit their needs. High-tech manufacturing is developing at a ferocious pace. As their products advance, so do the components they require, and this means relying on suppliers who can rapidly produce new products to exact specifications, not to mention relying on a workforce capable of altering their work without losing momentum. Companies that specialise in this are likely to rely on quality production and flexibility for their competitive advantage as opposed to price, and this gives domestic production an edge. High-tech manufacturers usually demand high-quality components to ensure the overall superiority of their goods, such as in military hardware manufacturing. Given the rapid change in these supply chains, a low-cost, high-volume producer's business model holds them back and they have little ability to ensure this quality or to develop their product according to rapidly changing specifications, rendering offshoring even less attractive. British companies, however, are well placed to supply this, combining their knowledge of domestic customers with experience to allow greater flexibility in production with the client in mind.

Textiles

As already discussed, the textile industry is one where products are often labour intensive, and so one would expect the majority of clothes meant for general consumption to be produced in places where labour is cheap. However, this is less true than might be assumed. The WTO's statistics on textile exports show that China accounted for 30.7 per cent of world trade in this sector in 2010, up from just 10.4 per cent a decade earlier. However, the EU-27 still produced 26.8 per cent of global exports in 2010, although this was down from 36.6 per cent in 2000. Most of these European exports are consumed by EU countries, with only 8.3 per cent being sold to other nations while China only exports a fifth of its products to Europe. The higher production cost EU is still the second-largest manufacturer of exported textiles, and it would be reasonable to have assumed China and Asia as a whole (26 per cent of global exports in 2010, with 4.4 per cent to Europe) would export much more to Europe.

Fashion changes constantly, so one season's must-have items could sell out one month and languish on shelves the next. This effectively undermines the high-volume economy of scale that would make Chinese production attractive. For manufacturers, this means instead that the lead time between order and delivery is key. There are still many cases in which basic, labour-intensive processes are performed via low-cost outsourced contracts, such as sewing or dying, but this suffers from issues with dependable quality as the separation of contractor and customer makes inspection and the maintenance of standards rather difficult. Many companies have onshored at least some of their operations to overcome the problems. Extensive research on this trend was carried out by the University of Aquila, which summarised the benefits:

> Providing the firm with a capacity buffer in order to respond in a proper manner to unexpected peaks in the demand and unforeseen variations of supplier lead time, thus improving the system's responsiveness; applying a stricter quality control to high value products; possibility of setting up a line for small sized pilot production runs for new products development as well as to satisfy customised requests.

The competitive advantage of these companies is their reaction time. This overrides the advantages of offshoring in this high-volume, labour-intensive industry. European and British factories have been upgraded with the latest technology to allow a just-in-time production model.

Supply chain resilience

A supply chain is only as strong as its weakest member, so its resilience is a critical issue for many manufacturers. As increasing numbers of supply chains have been offshored to a greater or lesser degree, the risk that there will be a disruption to the chain has also risen. Industrial action, war and natural disasters are all capable of undermining a global supply chain's ability. One way round this has been the onshoring of critical companies and activities, to protect them from international vulnerabilities and for those that can, shortening the supply chain can be a huge advantage. The resilience means all the involved companies can better mitigate against the risk that their supply chain will be disrupted. Indeed, the latest EEF survey found that during the recession, 60 per cent of UK companies were concerned about the vulnerabilities of overseas suppliers, compared to 20 per cent worried about domestic suppliers. As a result, two-thirds of companies have re-evaluated their supply chains to minimise these risks, and some brought production home while others began to buy components from local firms where possible.[7] There are further companies who would buy British if the parts they needed were available from UK suppliers. Increasingly, British offshorers are aware of this latent market, and repatriate production to take advantage of the demand.

Inter-British supply chains offers customers much greater flexibility in production and relative safety from a resource drought. By manufacturing the components themselves, or through a trusted and reliable domestic partner, businesses do not have to fear that their components will be subject to a delivery delay that will pass up the supply chain and annoy their customers. Tony Caldeira, of the eponymous company was motivated to move production back to Merseyside because 'we wanted more flexibility within our stock and we wanted to reduce our lead time.'[8] The importance of reliable delivery cannot be understated. In a 2007 survey, high-tech companies in particular stated that logistics was one of their top three sources of competitive strength and stressed that this would be even more important in the future, with the strongest responses from the UK manufacturing stalwarts of the automotive sector (85 per cent) and the electronics sector (72 per cent).[9] This is unsurprising. High-tech manufacturers often work along the principles of 'just in time' manufacturing, meaning that suppliers of parts must be able to change the speed of supply rapidly according to their customers' need. Having geographical proximity between supplier and customer optimises this ability. There is little flexibility when shipments have to be ordered months or weeks in advance from China and many companies would rather pay a premium to British suppliers that can harmonise with their supply needs.

In addition, for multinational companies having a plant at home and a plant abroad is the business equivalent of an 'heir and a spare'. It mitigates risk and ensures that production is still possible even if conditions in one area become very difficult. This is increasingly important for supply chains relying on just-in-time production methods: Toyota's profits collapsed by 75 per cent in the wake of Japanese earthquake and tsunami of March 2011 and it took the best part of the year to get Japanese production back to full capacity.[10] Because most of its parts were made in Japan, there was a severe shortage of Toyota components around the world and the assembly line in Derbyshire was forced to cut working hours for its 2,600 staff as a result. In the wake of the disaster, many Japanese firms have shifted their sourcing away from purely Japanese suppliers, to others, including UK-based ones, to prevent this happening again.

Anglia Components and United EMS

In November 2011, UK component distributor Anglia Components signed a component supply contract in excess of £3 million with United EMS, a UK electronic manufacturing services company. The awarding of a volume electronics contract to a UK manufacturer rather than their Chinese competition goes against the grain of the still predominant trend of offshore outsourcing large orders of electronics parts. LPRS, a British supplier of short-range radio devices has similarly flouted outsourcing conventions, creating a UK based supply chain in which Chippenham-based Danlers manufacture its easyRadio products, London's Transonics provide procurement and kitting services and Leicester-based Lewmax programme parts of its radio modules. The LPRS local outsourcing strategy was claimed by managing director John Sharples to be motivated by a need 'to find a more efficient way to run our business'.

These electronics firms have seen outsourcing to local contract manufacturers as a way of streamlining and eliminating the logistical problems of overseas production. Or as Tony McFadden, director and co-owner of United EMS, has more bluntly put it: 'there is an opportunity, a gap in the market to work with those people who have had headaches manufacturing in the Far East'. Electronics is an ideal sector for onshoring, where wages constitute a relatively small percentage of total production costs and in which logistics costs and issues, such as shipping time and distance, are critical.

United EMS has become the UK's fastest growing electronics manufacturing services provider by focusing on creating cost-effective supply chains for its clients. This means offsetting the cheaper price of components in China. McFadden claims that providing an effective UK-based supply chain depends upon the adoption of a different approach to costs and the supply chain to that typical in UK-based manufacturing businesses. United EMS' lower overhead cost model was achieved through investment in production equipment, obtaining necessary certifications and cost-effective staffing levels. These 'lean' production techniques enable manufacturers with UK-based supply chains to remain competitive by offering value for money as well as the prospect of avoiding logistical headaches.

Healthy working relationships are easier to maintain between firms with British production. Information can pass between supplier and client more easily and, as Betafence found (see p.32), there is clearly a demand for more relationships of this kind. This is especially important during the strain of recession, which can threaten to topple entire supply chains. While this risk is certainly not constrained to global supply chains, it can be somewhat mitigated through the negotiation and compromise only possible between firms in the same locality. For instance, trade credit insurance covers firms in the event that the customer does not pay for supplied goods, but the premium has increased rapidly in cost as insurers struggled with the economic downturn. With insurance now often too expensive for companies, manufacturers have increasingly relied on trust and openness as the means to ensure payment and avoid the cash-flow problems a disruption would create. Companies are most willing to engage in this if the supply chain is domestic, as dialogue can be opened more easily and reputation matters more.

The supply chain is not a one-way system and British offshorers have much to gain from onshoring beyond the potential for increased custom. Britain still produces large volumes of machinery and equipment used by other companies, including offshorers. If the firms who buy this equipment and sell to the domestic market onshored, there would be the double benefit of being close to their suppliers and customers. This creates good opportunities for efficiency improvements and innovation. A customer's demand for new products could highlight an opportunity for this and then the equipment supplier could rapidly enact the improvement for the company. This also creates the opportunity to form reciprocal relationships.

There are additional benefits for the companies and the supply chains as a whole from choosing to onshore. Geographical proximity also means that firms are likely to share similar values, which is important in the customer-supplier relationship. Chinese production has a high 'psychic distance', which means that the differences in culture throw up potentially unforeseen issues, such as ability to value independent action.[11] These affect the ability of foreign plants to react to the exact needs of the British firm, especially where innovation is concerned. Bringing production home often eliminates these issues entirely.

And the manufacturing that won't

There is no situation where it can be expected that all or even the majority offshored manufacturing will return to the UK. Conditions will never favour this. An obvious reason would be that many British companies want to supply the Asian markets, which are well known to be growing at a furious rate. By 2015, it is expected that Chinese disposable income will grow by 230 per cent, to $5.57 trillion. The burgeoning Chinese middle classes are increasingly aspirational and heavy consumers of all sorts of goods and by 2016, there will be 90 million households earning over $9,000 a year, putting them firmly in this category.[12] In India for example, while only 5% of the population was categorised as middle class in 2005, the National Centre for Applied Economic Research claims this will increase to 20% by 2015 and 40% by 2025.[13]

Between that and the need for increased amenities such as healthcare and housing for a rising urban population, there are huge opportunities for British companies to make large profits through offshoring, either directly by supplying the market with their goods or indirectly through supplying Chinese manufacturers with intermediate products.

It makes a lot of sense for these British manufacturers to make their products for these direct and indirect markets within or near to these far-flung markets. This gives a much better ability to tap into these growing markets and create an awareness of the product, which is highly advantageous for their future success. These companies sometimes operate dual production facilities in the UK and abroad, allowing them to retain full control of overall quality and the production process there as well as through 'offshoring insourcing'. It is usually only large companies that have the luxury of this option and when they do, the advantages of offshore production begin to pick up again. With full supervision, reliability and efficiency problems can be ironed out and labour cost savings are not dwarfed by shipping costs. However, it should be borne in mind that small and medium-sized companies rarely have the resources to operate multiple production plants. Given that they rely on outsourcing their manufacturing to foreign companies, they are far more likely to consider returning production to the UK.

For example, JCB built a factory near Shanghai in 2005, in part to supply Chinese demand for their products. While the company still retains production in various UK-based plants, it makes sense to produce the large machines for

Asia in Asia, given the shipping costs, and goods destined for European and other markets in Britain. For British companies in this situation, having foreign production has proved a lifeline during the recession. JCB struggled hard to continue producing a profit in 2009 but still managed to do so, in no small part owing to growing demand in China and India at a time when demand in Europe continued to flatline. As far as companies large enough to take advantage of this are concerned, this is the best of both worlds.

Companies specialising in the supply of components to the offshored plants of other firms will have little reason to return home, other than through a wider repatriation exodus. The Weir Group, a large engineering firm based in Glasgow moved 90 per cent of its production overseas to be near its customer base, and it simply makes no sense for them to move business back home.[14] Norman Hay, a manufacturer of coatings for the oil and motor industries is another such firm. In an interview with the *Sunday Times*, its owner explained:

> Ten years ago, a third of our auto sealants business was in the UK. Now it's tiny, a few percentage points. We follow our customers, which is why we have a facility in Dalian, China. We recently set up a new site in Malaysia.[15]

These companies effectively follow the same rule as those producing goods abroad for foreign markets.

Goods produced in large volumes will also remain offshored, if labour is a moderate component of the total costs. The need for proper infrastructure will tether companies producing these goods to China, despite lower-wage cost countries existing. It will be the manufacturers of mass-produced labour-intensive goods, such as high-street clothing or shoes, that will emigrate from China and remain offshored in the cheaper locations. Nike for instance has moved much of its production out of China, which used to produce much of its wares and into Vietnam, which now produces most of the company's shoes.

The evolution of British manufacturing

Beyond their position in the economy, the overall operating methods of British manufacturing companies have changed in a way that encourages onshoring. They rely on increasing levels of automation and the fusion of actual production with additional services, which gives them an edge on the competition. As is fairly obvious, Britain is not a low-cost country and its manufacturers cannot out-compete foreign rivals via price cutting. The steel industry now found in the UK is completely different to the steel industry one would find in China. British production of this basic product is based on quality, lead time and innovation: the ability to make things in a way no one else can. This edge derives from the highly advanced machines used, which cannot be found elsewhere in the world. British manufacturing's reliance on high-quality production is a given. For most firms, price cutting would erode profits and restrict investment capability, pulling firms into a slow downward spiral. Little wonder then that this simplest of ways to respond to competition was rejected by a quarter of UK firms without hesitation, a further half had considered but rejected it and only 14 per cent actually implemented a price cut.[16] Instead, British manufacturers have turned to other strategies to evolve their businesses to compete in the 21st century and ensure retaining UK production remains viable. Quite what a company focuses on to develop depends on the sector they are in. In the machinery sector it is the design that matters the most, while those in metals are reliant on serving individual customer needs and ability to deliver reliably. It is these evolutions that the government must be aware of and encourage as a means to enable onshoring.

Increased productivity through automation

As in the US, it will be impossible for Britain to recreate the manufacturing workforce size it once had: there is simply no need for this anymore. For example, Nissan's Sunderland plant produced 271,000 cars with 4,600 employees in 1999 but in 2011 it made 480,000 cars with 5,500 workers.[17] The steady rise in productivity is down to the increasingly advanced machines used. Automated production has developed far enough that FANUC, a large Japanese manufacturer of industrial robots, is able to run production lines without any supervision for weeks on end. In the UK, and indeed, for the vast majority of companies worldwide, this is an extreme outcome rather than the future norm. Even if they are not required on the factory floor itself at all times, humans will be paid to oversee, maintain and instruct this automated equipment. This is what employees of manufacturing firms will be paid to do, rather than produce anything with their own hands.

However, this is not the case yet. For the most part, robots are not advanced enough to overtake the human capacity for final product assembly, as they are not yet precise enough and even in the industry most associated with mass production by robots, the automotive sector, final assembly is still usually done by hand. Of course, for SMEs, it is also often prohibitively expensive to purchase automated machinery and the workforce will still be an integral part of their manufacturing process. For the likely future, there is every reason to expect, then, that onshoring will lead to a rise in employment and increasingly automated production will augment the efficiency of employees and enable them in new tasks rather than replace them. The numbers involved in manufacturing will be smaller, but certainly still significant.

Nonetheless, it is this diminishing cost of labour as a proportion of production costs that will drive the repatriation of British manufacturing. McLaren, which produces carbon fibre bodies in a new factory in Woking, is able to keep this production in the UK because the process only takes four hours, when just a few years ago they had to be almost fully hand-made, consuming 3,000 man-hours, an incredibly expensive operation for UK labour.[18] This goes hand-in-hand with the ability of advanced machinery to change products to suit changing needs quickly and cheaply: all this needs is a skilled workforce capable of supervising the change. The *Economist* summed the situation up well, announcing, perhaps prematurely, a third industrial revolution:

> As manufacturing goes digital, a third great change is now gathering pace. It will allow things to be made economically in much smaller numbers, more flexibly and with a much lower input of labour, thanks to new materials, completely

> new processes such as 3D printing, easy-to-use robots and new collaborative manufacturing services available online. The wheel is almost coming full circle, turning away from mass manufacturing and towards much more individualised production.[19]

For some markets, it is this move away from high-volume economies of scale as the key to success and a shift towards customisable production that will pull industries home just as much as higher costs push them out of China. For other markets, this will be less important, such as in the automotive sector.

Service provision

There has been a general shift in most British manufacturing companies who now see themselves not just as providers of goods but providers of services as well. Many firms have begun to blur the line between the supposedly distinct economic realms of manufacturing and services in a bid to create a new competitive advantage in light of greater global competition. This phenomenon has been slowly increasing in the UK, with 28 per cent of companies offering this by 2000, a faster rate of uptake than in many countries like Germany and France, and by 2007, over 50 per cent of respondents to EEF's survey said developing services was a key strategy they were using to find a new competitive advantage.[20] Service provision is a useful 'insurance policy' for British manufacturers, as even at times of decreased demand for their goods, companies can sustain themselves by implementing their services for a continual revenue stream. The hybrid model is more attractive to manufacturers' customers, as they avoid the hassle of having to find a service provider and this compensates for the higher costs they pay for British products. This additionally makes production far more exportable than would be the case if just the product was supplied. In total, it is estimated that over half of the roughly three million workers employed in manufacturing are actually involved in things other than the production but vital to maintaining a competitive edge, such as R&D and services: manufacturers are now more than just their products.[21]

Rolls-Royce is a good example of this trend and is a company that pioneered the augmenting of initial profit through selling a good with follow-up care. Its engines are a one-off source of cash but the maintenance that goes into them over their lifetime is a continual revenue flow. This also benefits suppliers of Rolls' components, who will know that their products will be required for many years to come, which helps drive down the production costs of the initial engines. Many firms have now integrated this fully into their business model and like Rolls-Royce, most of these are large firms which are better able to devote manpower and resources to creating new service offerings with relative ease. For instance, the shipbuilding and services group, VT, now makes as much profit from servicing their ships as they do from the initial purchase.[22] Now, over 65 per cent of large British manufacturers offer services according to EEF, as opposed to around half of small and medium-sized businesses.[23]

Continuous Manufacturing

Continuous manufacturing has great potential to drive onshoring. This process, where the production of a good happens at once, aims to maximise efficiency while reducing cost and the number of defective products. Rather than shipping raw materials and parts from one location to another to eventually assemble the end-good, continuous manufacturing offers a way to do all of this at once, reducing the supply chain length and improving logistics. While still a relatively pioneering concept, there has already been great progress in the pharmaceutical sector, which contributes a £7 billion positive trade balance to the UK and employs 72,000 people in Britain. Rather than having to use various plants around the world, the Swiss pharmaceutical company Novartis, which has a UK base of operations, has developed a production line that inputs raw materials and outputs finished pills. While still five to ten years away from commercialisation, it has so far reduced discrete operations in producing drugs from 22 to 13, and, discounting time spent freighting components, reduced processing time from 300 hours to 40. Given that it can work in smaller batches and is itself much smaller than conventional equipment, this production line would allow a good degree of import substitution.

FeONIC

Whilst some manufacturers sell long-term service contracts, guaranteeing maintenance and after-sales care along with the product, others design bespoke products tailored to the customer's needs. Among the latter group is FeONIC, a company which originated as a spin-off from Hull University and claims to offer its clients solutions to 'acoustic challenges'. As specialists in the design and development of magnetostrictive audio products, FeONIC provides highly intelligible wide bandwidth sound by vibrating structures including floors, windows, walls and even ships. These materials then emit a sound understandable by humans, thereby avoiding the need for conventional speakers. The smart material used in its products was originally developed for sonar devices by the US Navy.

FeONIC's USP revolves around the concept of 'responsiveness'. The company works closely with clients, offering a bespoke design service, often for unusual applications of its technology such as its 'Whispering Window' retail displays, which play advertising messages as customers pass by the display. FeONIC has adapted its technology for a range of purposes from high-quality railway public address systems to waterproof yacht entertainment equipment and this is a strategy with which overseas suppliers cannot compete. Manu-services firms like FeONIC capitalise on customers' increasing expectations of quality of experience as well as of product.

It is perhaps surprising that so many SMEs can offer services, but this is certainly something to applaud. While many small suppliers have forged a niche for themselves by supplying larger companies with specialised components, requiring nothing post-production other than goods logistics, many other suppliers are not able to entrench their advantage this way. Service provision offers them a solution. Gerd Bender, of the University of Dortmund, described an Austrian rail-track manufacturer who adapted to the demands of its customers who 'tend to ask for system solutions rather than simply tracks'.[24] The firm had developed the world's longest piece of rail track, but this was in itself less attractive to customers compared to conventional track given it was difficult to handle. The natural and most successful solution was to integrate the track laying service into the process of supply, which the firm did through inventing a machine capable of dealing with the long rails. The customer could therefore benefit from all the advantages of the long track without any of the worries otherwise involved. Even for those within niche markets and a comfortable level of security, their usually quasi-unique position means they are already often consulted by others in their sector. This supply of knowledge could become increasingly commercially viable to such firms either directly, through consulting, or indirectly, via innovation in supplying system solutions.

Ken Coutts and Robert Rowthorne of the University of Cambridge investigated the potential for British firms have to develop the hybrid production-service model further, and found that, 'it would be difficult to conceive of a viable industrial policy for manufacturing that did not also involve knowledge-intensive services'.[25] The expansion of the manufacturing sector in general will mean a rise in demand for the services and system solutions that manufacturers can offer. They suggest that, 'access to this market would enable UK service providers to benefit from economies of scale and develop skills which can be exploited in export markets'.[26] It is likely that many offshorers will be attracted to the hybrid model and will return production to the UK to implement it. As foreign competitors begin to develop their advanced technology further and increase competition on a product quality basis, it will be the provision of service that will set these British companies apart. To maximise the ability to provide services of just as a high a quality as their products, many will base themselves near to their customers and this means onshoring, given that Britain is still a stronghold of the sort of high-tech company that would want holistic solutions. This would give onshorers an edge that their geographically distant competitors will be unable to offer.

3D printing

3D, or additive printing, is a relatively new technology that has significant potential to boost UK manufacturing. The process, whereby custom-designed parts are 'printed' by constructing them layer by layer, allows very complex products to be created from many different materials without the expense of tooling and without much waste, as the machine only uses as much raw material as is required. The process was originally developed to produce prototype products, but, with technological improvements, it is now possible to create finished goods with the printers. As 3D printing does not benefit from any economy of scale, it is most useful for producing goods in small quantities and with mass customisation. This could make production of low-level, low-cost goods in the UK a real possibility again, as it has the potential to cancel out the labour cost advantage of emerging economies. This is highly useful for British businesses. Many companies find that when looking for components, the low-cost offering from offshoring are second-best, and while viable for use, are not specifically what they would like to buy. 3D printing offers an escape route, allowing them to print the parts themselves, or utilise the expertise and volume potential of a 3D printing supplier. Either way, there is no advantage in offshoring this production. The cost of the materials and printer will be the same wherever it is set up and, given that labour cost is minimal, shipping costs will make British production most attractive. Additionally, 3D printing plays to British strengths by allowing the production of high-quality products that are simply too difficult to produce by traditional manufacturing methods. For example, the UK firm 3T RPD has printed a racing car gearbox with smooth internal pathways for hydraulic oil instead of the usual drilled out right-angles. As fluid flows better around bends than right-angles, this unlocks efficiency improvements that were previously unreachable.[27]

Britain is already developing its 3D printing capabilities. The Engineering and Physical Sciences Research Council has a Centre for Innovative Manufacturing in Additive Manufacturing in Loughborough, which aims to become a world centre for the technology. As part of this it has secured £4.9 million from the Research Council and a further £3.2 million from 16 multinationals and high-tech SMEs. The aim of the Centre is to harness the potential of 3D printing and produce ready assembled products in different materials that would be used in a variety of industries. For a different market, the University of Bath is developing a low start-up price machine, which would produce goods quickly. The most well-known output of the University, and brainchild of Dr Adrian Bowyer, is the 'RepRap', which can produce a range of everyday plastic goods such as coat hooks. More intriguingly, it is also able to reproduce almost all of its parts and can essentially print itself.

While 3D printing is not yet established or viable for most British companies, it certainly has potential, and could be instrumental in preventing a new wave of offshoring in the future. At the moment, 28 per cent of the money spent on 3D printing is for finished products, with the remainder still being used to build prototypes. However, it is expected that this will reach 50 per cent by 2016 and 80 per cent by 2020.[28] If Britain can stay ahead of the curve with 3D printing, it could reap long-term rewards.

4

How should Britain encourage onshoring?

Why we need to encourage onshoring

Despite the pessimism generated by the recession, there are reasons to view the British manufacturing sector in an optimistic light. The monthly Purchasing Managers Index, run by Markit and the Chartered Institute of Purchasing and Supply records the manufacturing industry's condition and reveals the sector's confidence. A reading of over 50 points is considered a sign of growth and as of March 2012, the value was 51.9, a ten-month high which is a good sign post-recession.[1] Even the recession itself did not greatly damage manufacturing confidence, with shrinkage reported only from early 2008 to a low of 34.9 in early 2009. Confidence rapidly increased again so that positive growth over 50 points was recorded by Q4 2010. Overall, growth has been healthily above 50 since 1997. The rise in recent months has been put down to an increase in output and new orders from home and abroad.[2] With the manufacturing sector growing again, there is no doubt that the Government should be trying to encourage British manufacturers to return their production home or outsource to UK companies. Research by Coutts and

Rowthorn in 2010 suggested that a rise of 10 per cent in goods exported and a 10 per cent decrease in goods imported would contribute £45 billion to the UK economy.[3] To put this in perspective, this is almost two-thirds greater than the 2010 Current Account deficit of -£29 billion.

On the whole, the UK is already seen as a good place to do business by companies. The World Economic Forum's Global Competitiveness Index for 2011-12 rates Britain as the tenth most competitive economy, up from twelfth place in 2010-11.[4] In contrast, China came 26th, up from 27 the previous year, while India came 56th, down from 51st previously. Britain was noted particularly for its efficiency factors, which placed it fifth in the world. However, the government and civil service came out of the survey less favourably. 'Inefficient government bureaucracy' was claimed by 13.5 per cent of respondents to be an important barrier to doing business in the UK, and the third most frequent response, a rise from fourth place in 2010-11.[5] In part, this poor outcome is a response to the lack of understanding the government has shown towards the manufacturing sector and its needs. It has pressed ahead with its own political conception of what the country needs to revitalise the economy but has failed to actually provide the support industrial companies need if they are to repatriate their production. This is echoed in EEF's survey of manufacturers, which found that among firms already manufacturing in Britain, over 40 per cent of companies felt the UK was a good place to do business with 20 per cent strongly agreeing, but 37 per cent disagreeing. Clearly, there is still much work to be done.[6]

It should be remembered that the UK's attractiveness can only be seen in relation to other countries' business environments and these other location opportunities. While bureaucratic inefficiencies, logistical and other problems are often rampant in offshore production locations, this is not something that will last forever. China is constantly trying to improve its attractiveness and has built 42 airports in the last decade and has a further 55 planned over the next eight years.[7] In the same time period, the UK has deliberated about building one single extra runway at Heathrow: Britain cannot afford to sit still and expect onshoring just to happen. To retain our high position in the competitive index, and more ambitiously, to improve it, new policies are required. The Boston Consulting Group's report on the US insourcing trend currently sees little likelihood of Europe (and therefore Britain) experiencing the same repatriation possibilities as the US. Indeed, BCG suggests 'manufacturers from Western Europe... could begin to establish more production facilities in the US to serve domestic and European markets'.[8] Much of this is blamed on European wage costs, and, adjusted for productivity, Chinese labour will be just 38 per cent of average European labour in 2015, which it believes is not enough to create a 'tipping point' akin to that of the US.[9] This is a rather pessimistic view of the effect of current policies, and should act as a wake-up call for the Government.

In trying to encourage onshoring, the Government has to realise that this is not as simple as it looks. It is not just a case of trying to woo individual companies back to the UK, but involves whole supply chains relocating. As already discussed, many SMEs supplying larger companies originally decided to offshore their production not so much because of the cost advantages but because they were following their customers. Component manufacturers found it much cheaper to supply their products in the regions where they were required, and their potential motivation to return home will be driven by similar requirements. They will repatriate if their customers do so, but not every company, that would like to repatriate production, will be able to if left to themselves. This is especially true of SMEs, for whom the costs of offshoring in the first place were too high to allow production to be moved again in a short period of time. Without government assistance, they will have no choice but to continue offshore production and with all the discussed rising disadvantages, this might undermine their competitive edge in the long run. Given the additional contribution they would be making to the British economy, it is right to offer them a lifeline in return for a long-term commitment to establishing production facilities here.

Many of the below recommendations will also encourage foreign investment in UK manufacturing, either through new plants, extensions of existing capabilities or even the prevention of offshoring. This is something tangential to this discussion but still pertinent: the bulk of all foreign direct investment goes to manufacturing. If the Government fails to improve Britain's attraction in these key areas, the country could not only find itself losing this foreign investment, but also more British businesses. The onshoring trend will only last for as long as the UK remains attractive and should the country fall behind, offshoring will again become the norm.

Support for manufacturers of intermediate goods

The Government's plans for increasing manufacturing output are certainly noble, but will be less effective if they continue to focus purely on high-tech components and finished goods. Many of these companies require parts from other firms, and without support for British manufacturers of intermediate goods, this will lead to a steady rise in imports.

While visualising what high-tech manufacturing is might be easy, the actual sectors the government should be fighting to onshore are the more intangible and general 'high value' ones, which are not necessarily those producing advanced or finished goods. This 'high value' is calculated through the valued added contribution made by industries to the economy. For the UK, the results are somewhat surprising, as seemingly basic products are a real British strength. In 2007, low-tech goods accounted for 37 per cent of manufacturing value added, while high-tech ones contributed just 17 per cent. The UK has a fairly middle of the road ratio of high-tech to low goods output, compared to countries such as South Korea.[10] Clearly, in Britain, high-tech does not automatically equate to high value, and this is something the Government must realise. The Department for Business, Innovation and Skills has focussed its support within high-tech provision and the Chancellor's Budget 2011 speech highlighted only his intention to provide for high-value and advanced manufacturing, with little reference to other manufacturing sectors.[11] This means that there are plenty of low-tech offshorers who are ignored by current policies aiming to make the UK more attractive as a place to do business.

Promotion of high-tech manufacturing is undeniably positive and indeed the future of British manufacturing in terms of exports is and will continue to be high-tech dominated. The UK has a significant competitive advantage compared to many countries in terms of its knowledge base, technical expertise and manufacturing equipment: these must be used to entrench our global trade niche. As such, the concept of the 'knowledge economy' has been widely publicised as the eventual shape of the UK's economy which involves having a highly-trained workforce and the free flow of information. However, a policy aiming at fostering high-tech growth must come in the form of a general industrial policy that encompasses all manufacturing. This is because almost all high-tech manufacturers utilise components made by low and medium-technology suppliers. While many of their parts are already sourced from British companies, a generally inclusive industrial policy would increase this ability through onshoring.

This theory was confirmed by the research group 'Policy and Innovation in Low-Tech' (PILOT), which conducted a Europe-wide study of low-tech industries from 2002-05. This was the first in-depth investigation into the assumed decline of LMT in 'post-industrial' societies, but it found that the knowledge-based innovation in LMT manufacturers was the manufacturing sector's greatest intangible asset and the source of long-term potential growth in Europe.[12] This was reliant, however, on sustaining existing industries through innovation, as opposed to the fostering of new ones, which is more akin to the current British government's policy. The PILOT survey found that low-tech firms involved in the seemingly basic manufacture of products such as 'wooden boxes' or 'simple metallic parts' had the potential to grow very fast because of the advanced production processes on which they rely.[13] Such growth in basic commodity production is a result of the fact that they are widely used in many areas of industry. Their success is their simplicity, but this is not to say that they would continue to be offshored if the government's support for them increased.

Improve skills

As discussed, the increasing reliance on automated production and computer simulation of products means that the average manufacturing worker requires higher levels of education and a broader range of skills than before. While 'traditional' skills such as engineering and science are still very useful, if not critical, there is a rising need to merge these with expertise in new ones such as information technology, to harmonise with the production processes. As the example of America shows, encouraging onshoring will not be driven by a rise in the number of apprentices or graduates alone, but by the quality of education on offer and fusion with actual work experience. This is as true for low-tech sectors as it is for high-tech ones, and the overall proportion of degree holders in manufacturing have jumped from 8.9 per cent of the manufacturing workforce in 1994 to 16.4 per cent by 2006.[14]

There is a dire need for an improvement in basic skills provision in the UK as a 2005 report by the OECD stated that British average worker skills are lacking: 16 per cent of the working population lacked basic literacy skills and 21 per cent lacked basic numeracy ability.[15] While this situation has been improving, sufficient progress is yet to be made and this lack of skills would act as a drag on national ability to innovate, no matter how much funding the government channels into higher education and R&D.[16] Importantly, the unacceptable standard of education has moved on from just being an academic problem. A study from 2010 has also shown that British hiring firms in the industrial sector are still frustrated by school leavers' lack of skills, with 11 per cent believing school leavers are not equipped with the required numerical ability.[17] There is a simple need to raise the ability of the lowest skilled workers as well as improve them at the top, to fully maximise the potential of the manufacturing sector.

The UK is not managing to keep hold of its lead in global education rates and has continued to fall down the rankings of OECD countries – not because the level of national skills has fallen, but because so many other countries are improving theirs while ours stagnate. In the 2009 PISA assessment, which is conducted triennially by the OECD by testing 15-year olds' abilities, Britain came 26th on reading ability, 28th on maths ability and 16th on science ability, placing it 25th overall, a middling position for a developed nation.[18] This is a clear decline on the results three years previously, when it came 17th on reading ability, 24th on maths ability and 14th on science ability.[19] Shanghai came top in all three test areas in 2009.

This decline, while trying for companies, is not insurmountable for all of them. Companies that rely on computer systems and software to produce or regulate production of their goods have an advantage as they can simply let their trainees learn on the virtual job, by creating simulations of their programs. This is far less time consuming and wasteful of valuable, experienced worker time than teaching them on actual equipment, as would have been necessary a few years ago. However, firms that enjoy this luxury, and have the spare staff to devote to training are not widespread. Moreover, many firms that relied on traditional, more expensive training, tried to avoid the recession biting into their bottom line by cutting back on the training they supplied to their workforce. While this has kept many in business, it means that as we are now coming out of the downturn, their ability to increase efficiency is somewhat stunted until this training and investment takes place and the longer this does not occur, the further behind these companies will fall and the less likely they will ever fully recover. This is where the Government could step in, by providing quality further education that would allow employees to grasp industrial training more easily. By doing so, the Government would enable onshoring in a much more systemic way.

Advanced qualifications

The size of the highly skilled workforce onshorers require is being limited by the Government's policy of pushing young people through tertiary academic qualifications and viewing advanced vocational training as secondary to gaining degrees, whatever the quality. In recent years, this has been an increasingly maligned area as both EU and self-imposed targets for the number of university graduates has taken priority in the tertiary education focus. The EU aim of ensuring that 'at least 40 per cent of the younger generation should have a tertiary degree' promotes education based on quantity rather than quality.[20] In the UK, the solution was an increase in the availability of two-year degrees, a concept Vince Cable supported, which was launched in 2010.[21] Two years on and the response has been a negative one from businesses. Issues cited have included a lack of maturity in two-year graduates, not enough transferable skills because no extra-curricular activities took place and not enough development. This attempt to maximise the volume of students while minimising the costs of doing so will only serve to undermine the initial well-intentioned aim of increasing the educated workforce, especially in STEM subjects where two years is simply not enough to ensure graduates can maximise their skills in the workplace. A 2010 survey of industrial companies found that of new jobs offered, 53 per cent were given to those with previous work in the sector while 17 per cent of employers felt new entrants did not have the practical expertise they required.[22]

This is a critical issue that increasingly stunted British manufacturing in general: graduates might have had the skills required, but this alone does not make them employable. EEF has done its own research into this issue, and found that if high levels of skill are required, then companies will seek out those with PhD or postgraduate qualifications rather than graduates with just a degree. This is partially because they are older and more mature, but also because they have the really strong technical expertise companies increasingly require, that allows them to be placed in any division of the company. However, it could be argued that the need for PhDs is a symptom of the poor quality of other graduates. If this general standard was raised, then the expense of having to provide costly doctorates could be minimised.

A 'Catch-22' situation is emerging where graduates require direct experience of industry in order to be employed, but they can only gain this involvement through having a job in the first place. This is the product of the relationship between STEM degrees and apprenticeships becoming very polarised, with the former relying on teaching theoretical skills rather and the latter entirely practical and using theory only insofar as it is required to aid physical training. Oxbridge for example teaches engineering without any compulsory placements during the four year course while other universities have an 'optional' year's placement in industry. The responsibility for gaining experience therefore falls to the student, not the university and there are clear financial incentives to avoid spending an extra year gaining practical experience. In addition to the cost of student living, tuition fees are still payable during that period at up to half the usual amount. It might appear obvious to suggest that these placements should be incentivised for students, but, at present, they are penalised. Removing the obligation to pay tutorial fees for the year out would be a good start and the unhelpful tertiary education targets should be abandoned.

Attitudes to vocational training

In recent decades, a culture has developed where vocational qualifications are seen as inferior to gaining an academic equivalent and this is true not just in political circles, but society more widely. Unless manufacturing is to stagnate and only attract second-rate workers, this ethos needs to rapidly change, with vocational courses given the respect they deserve. Informal investigations into the issue by trade bodies have found that the pressure to choose academic qualifications comes mainly from teachers, career advisors and perhaps most importantly, parents. This is in part a product of the continued assumption that industrial employment is low-skilled and low-paid, but another important factor is the historical treatment of manufacturing. Particularly in regions where industry was once strong, many parents and other influential people lost their jobs, leading to their aversion towards the new generation entering the same sectors. This is important, and a survey by the Engineering and Machinery Alliance found that a 'better image for manufacturing' was rated as one of the top three influencing factors of success by respondents.[23] Overall the assumption is that smart people go to university: it is the rest that look for apprenticeships. This is not a healthy attitude to cultivate.

The lack of official championing of apprenticeships suggests, even if it is not true, that the Government considers apprenticeships to be second-rate qualifications. It needs to clearly state that it considers them to be equal to a degree but inherently different, and so it should not keep trying to compare or merge them. To label an essentially vocational programme at a university a 'degree', and then remove the 'hands-on' modules is to undermine the strengths of both academic and practical learning and perpetuates the second-class nature of apprenticeships. The Government's expansion of funding for increased numbers of higher apprenticeships and incentives for SMEs to take on apprenticeships is incredibly welcome, yet more needs to be done to develop the same institutional respect for these practical qualifications that degrees already command.

A more insidious issue with the degree preoccupation is that government enthusiasm for vocational courses has decreased, something at odds with the demands for it. Regardless of Vince Cable's announcements in November 2011 regarding the increasing number of employers offering apprenticeships, it is questionable whether the scheme as a whole will receive sufficient support to reach its potential. Overall, the demand from would-be apprentices vastly outnumbers the number of places available. When recruiting for the annual intake of new recruits, BT was already swamped by August 2010, having received some 24,000 applicants for its 221 apprenticeship positions.[24] Clearly, there is a demand from many school-leavers for an alternative to a degree, but that enthusiasm is not matched by the government: from 2003 to 2009, the availability of 'engineering, manufacturing and technology' apprenticeships increased by only 3,900 places from 33,100 to 37,000.[25] The heavy limitations on choice have meant that most who want rigorous vocational courses have been forced to take an academic degree and go (often unwillingly) to university or alternatively to just start work, if they can even find a job. In manufacturing, this means they lack skills on which the production and innovation processes rely.

At present, many would-be onshorers will suffer the same fate as existing British SMEs when trying to find apprentices: while they might be world-class in their field, without a well-known name, no one wants to work for them. Some suffer further, because they are the suppliers to big brands and are hidden beneath them. It is the providers of the end product, such as Rolls-Royce or BAE, which are heavily oversubscribed as explained above, while everyone else suffers a chronic shortage of applicants. For instance, Midlands-based JJ Churchill is a world-class aerospace parts manufacturer, but because it is not the end producer in the supply chain, it struggles to fill its apprenticeships. This is a real problem, especially since many of those battling to find apprentices are in the same supply chain as the big names, so a lack of incoming trainees will have a knock-on effect on their ability to maintain quality production. A solution would be for the Government to create 'timeshare' supply-chain based apprenticeships, where apprentices spent time at all the companies in the chain. Given the majority of onshorers would be fitting into an existing supply chain on their return, this means they do not have to worry about struggling to find new recruits. The timeshare concept balances out the unequal distribution and would create apprentices with broader experience and awareness of the bigger picture. This is vital on all fronts, not only to maintain quality of product and service, but also to ensure these workers understand the needs of suppliers and customers and can better respond and innovate according to their needs.

Surgical Innovations

Surgical Innovations is a designer and manufacturer of keyhole surgery technology - a market dominated by large, often US, manufacturers. Surgical Innovations' success has been rooted in an unusual approach to its market. Traditionally, medical manufacturers have offered either disposable or entirely reusable laparoscopic (keyhole) tools but this company has built a fast growing business around the idea of making, significantly cheaper, equipment with both reusable and disposable parts, a concept which they have trademarked under the name 'resposable'.

Previously Surgical Innovations outsourced the manufacture of its designs to countries including China, Hungary, Morocco, Poland and Sweden. This wide dispersal meant senior managers were often away for up to a week and needed two days to recover from trips. According to the company's chairman, Doug Liversidge, this wasted time and the managerial dislocation impeded the company's progress. In addition, there were suspicions that in its Chinese operations, the company's IP and know-how had been given to local engineers to copy and that covert production of their goods was occurring. It was suspected that these were being sold at knock-down prices in Asian markets. Since 2008 the company has invested nearly £3 million in production machinery for the UK and today makes the majority of its products at its base in Leeds. Liversidge claims to have seen significant improvements in the quality of its manufactures since doing so.

However, Surgical Innovations quickly ran into difficulties as it proved difficult to find qualified staff to operate its complex equipment, which, as director Paul Birtles puts it, 'isn't just about pressing a button on a machine'. In order to overcome this obstacle, Surgical Innovations has taken on apprentices which is expensive but, Birtle argues, is 'the only way to get good people'. Each apprentice is trained on the job and attends college on day release for one or two days a week, at a cost of between £5,000 and £6,000 (on top of salaries) to the company. Therein lies the paradox: it is skilled manufacturing which offers the greatest window for onshoring to the UK and yet this process is hampered by the absence of a sufficiently skilled workforce.

An additional source of valuable employees is the adult apprenticeship scheme, which supplies many workers to the manufacturing industry. The scheme is crucial for getting the long-term unemploymed back into work and many manufacturers actually prefer using adult apprentices over younger ones. This is primarily because they are seen as more responsible, having a greater stake in working hard as they often have mortgages to pay off or families. Adults approach the apprenticeship with the knowledge that it could turn their life around. The demand for these adult courses is reflected in the statistics. In the year 2006/07 the number of apprenticeship starts for over-25s jumped from 300 to 27,200 the next year and reached 182,100 in 2010/11. This is incredibly promising, and adults outnumbered 19-24 year olds for the first time last year, by roughly 40,000 starts.[26] The rapid growth of over-25s apprentices shows that this group has been a relatively untapped but attractive resource. For a manufacturing renaissance to continue, funding for this group will need to be sustained.

More active intervention

Providing incentives to companies might not be the deciding factor when a company considers repatriating its production, but it can certainly ease the decision in conjunction with the 'push' factors from China. While some might feel uncomfortable with the idea of the state helping people financially, this should not concern the Government. There is a basic, latent support for manufacturing in the UK, and a 2009 opinion poll found that 57 per cent of Britons felt: 'the government should use public funds to help large manufacturing companies in trouble'.[27] More than this, it is clear that the money spent on helping manufacturers will return to the state in time, through corporation tax, income tax, VAT and other revenue streams. These measures should be considered loans and investments rather than funds never to be seen again

The Government currently intends to support specific industries such as low-carbon power generators. This approach is too narrow though, as the source of their competitive advantage may be sourced further down the supply chain at the point of interconnectivity. To overly define support risks failing to help the companies further down the supply chains, where the relationships become less of a 'chain' and more of a 'web'. The Government should be aware that unlike the producers of finished goods and high-tech ones in particular, the component manufacturers can supply very similar parts to a great number of other firms, particularly so if the products are low-tech. A producer of automotive castings for example can supply many car manufacturers, but also makers of heavy goods vehicles or earth moving machines. Even at a more specialised level, such as radar production, a firm could split their supply between military and civil contracts. This is increasingly the case as firms have been forced to shield themselves from defence cutbacks by entering into adjacent markets. One such British supplier of underwater electronic cables has dealt with the decline in defence contracts by moving into the offshore wind turbine market. Wind power relies on efficient cables to relay the generated power without loss, so this company was ideally placed with its previous experience to provide this. Britain's future manufacturing strength therefore relies on this ability to 'multitask' these various production processes. Policies aimed at actively intervening in industry at a national level need to take this into account, and aim to help manufacturers of all sectors.

In addition, it is very apparent that our rivals actively assist specific industrial companies and provide foreign firms with incentives to settle there. The US is an obvious example and Mars' decision to build the its new chocolate plant in Kansas (see p.79) was about more than just the attractiveness of the area. Importantly, both the state and local governments actively incentivised the establishment, by offering a package of $9.1 million's worth of deals. The

American Institute for Economic Research described the breakdown:

> $1.5 million worth of free land in an industrial park, $2.5 million to cover training and startup costs for its employees, and $1.7 million in workforce development funds. In addition, the Kansas Department of Transportation will invest $4.1 million for road construction and improvement to rail infrastructure. The Kansas Department of Commerce will provide another $1.85 million for further infrastructure development.[28]

Mars was hardly being left to make the decision on its own. This is exactly the sort of aggressive courting of businesses that the UK requires and once engaged in to great success. Margaret Thatcher might be remembered for her swathes of privatisation, but she very successfully secured foreign investment in the UK through attractive deals. When Nissan agreed to build a plant in the UK in February 1984, the government agreed to sell it greenfield land outside Sunderland for agricultural prices of £1,800 per acre.[29] This helped to ensure that the site was built in an area of high unemployment that would enrich the local area. As far as Thatcher was concerned, the government was not 'picking winners,' but guaranteeing employment for thousands. Flash forward and Nissan announced in April 2012 that it will build a new hatchback in Sunderland, creating 1,000 jobs to bring the plant's total workforce up to 6,225 employees. Indirectly, it is estimated that a further 3,000 jobs will be created in and around the North East of England as a result.[30] This 1980s investment has clearly paid off.

As well as offering general incentives to return production home, there should be specific funding to encourage onshore entrants into existing supply chains. The recession winnowed away at many supply chains, forcing many British companies previously able to access British components to turn towards offshore alternatives of dubious quality. While many industries have been able to survive this way, there is a great deal to be said for import substitution through the re-establishment of UK-based supply chains where possible.

Traditionally, the Government has also taken a vital role in promoting British goods abroad but this is currently under threat. The October 2010 Spending Review resulted in a 25 per cent reduction in the UK Trade & Investment's budget for trade promotion, reducing the worldwide publicity that British goods will receive. This is in contrast to Prime Minister David Cameron's own words: 'British business should have no more vocal champion than the British government and that's why I have put the promotion of British commerce and international trade at the heart of our foreign and economic policy.'[31] While Britain is cutting back on this national advertising, other countries are continuing to support their manufacturers. The Government will be sending a negative message to British firms and failing to supply the tools of industry that the EEF and BCC both say are expected and needed.

Import levies

There are reasonable grounds for the British government or the EU more widely to impose import duties on certain Chinese goods that benefit from underhand and morally dubious practices such as IP infringement or poor working conditions. This protects British companies retaining UK production and would encourage offshorers to return home, as they would be reassured that their business would not be undercut by rivals overly exploiting emerging economies. For instance, many energy intensive industries in China are easily able to undercut their British and European rivals owing to the fact that they do not have to pay much in the way of environmental levies. They are not financially penalised for their emissions or disproportionate energy usage, while European companies have to invest in expensive measures to reduce their emissions and increase efficiency. In the case of the Chinese aluminium industry, the US Commerce Department ruled that the sector was also receiving unfair subsidies.[32] This is having an effect on UK businesses and Britain has just seen its last major aluminium plant, in Lynemouth, Northumberland, close, at a cost of over 300 jobs. Instead, production flocks to the countries that offer cheaper operating costs.[33]

Imposing tariffs is not even a particularly bold manoeuvre. The US has imposed various import tariffs on Chinese goods to balance out perceived unfair advantages and most recently, it imposed duties on solar panels, which began in March 2012 at a rate of 2.9 to 4.73 per cent. It claimed that the subsidies Chinese panel manufacturers were receiving were far too large and drove down the average cost of panels by 30 per cent, as other manufacturers struggled to remain attractive. Earlier, in 2009, the US also imposed extra duties on certain Chinese tires that were felt to disrupt the market. The tariffs, of 25 to 35 per cent were allowed under the WTO 'safeguard' regulation, which aims to protect a country's industry from sudden floods of imports.

It would be acceptable to levy a duty on Chinese imports benefitting from this to protect the European industries acting responsibly and to take account of the externality cost of higher emissions. This will not only protect jobs, but will also ensure carbon leakage does not occur, as companies would gain no benefit from moving production of goods outside of the EU to areas with cheaper energy and fewer environmental costs. For maximum effect though, this would have to be implemented to include other attractive locations for energy-intensive areas such as Turkey and North Africa. Nonetheless, starting with China is a good way to balance out the greatest inequalities in the market.

Create confidence in the business environment

Confidence is a factor highly valued by companies but often ignored by the government. Without confidence of the future costs and benefits of British production, offshorers are unlikely to return production home, as the risk that the business environment will turn against British production again is too great.

Companies want certainty of state-related costs and benefits to allow them to plan ahead for the long term and this applies as well to the stability of the currency and wider economy. However, confidence in British government policy has been increasingly weak, and many companies have complained that the only surety they have is that costs are certain to rise in the future. Policy is too fluid to predict the future business environment with any clarity and this discourages companies from onshoring. One example would be the Lynemouth aluminium smelter, which closed in May 2012 at an initial cost of 323 direct jobs and a further 3,500 down the supply chain.[34] The official explanation from owner Rio Tinto Alcan was that 'energy costs are increasing significantly' and a 'thorough strategic review' were the cause of its demise.[35] Primary aluminium manufacturing has now been effectively offshored from the UK, with only a few small plants remaining here. While it is true that energy costs are increasing across Europe, they are not rising as fast or as unpredictability as those in the UK, which seems to have a new levy imposed almost yearly. Indeed, some countries have attempted to provide long-term price anchors, to avoid the same offshoring problem. In Iceland, Rio Tinto Alcan secured a 26-year electricity contract for its aluminium facility and will consequently invest $350 million to modernise the plant and increase output by 20 per cent.[36] This guarantee gives a level of forward security almost unheard of now in the UK.

Britain needs policies that create the opposite environment to that currently available in China: where there are uncertainties, the UK needs certainty. If companies are contemplating returning manufacturing to the UK, they will be doing so with the intention of keeping it there for a long time. These are exactly the sort of firms the government should be courting, as they will employ, produce and export for years to come. This means deploying long-term, clear and attractive policies that ensure the UK will reinforce competitive advantages over decades. Britain needs to make itself the permanent destination of choice for manufacturers, and avoid appearing as much of a fad location as China.

Improve innovation assistance

Britain's high-tech industry has weakened in recent years, in part due of the rise of China and the like, but also because the UK itself is losing its international appeal to these companies, who have been offshoring or, more frequently, are foreign companies pulling out of UK operations. As a whole, the British world export share of high-tech goods has fallen from 6.6 per cent in 2001 to 3.8 per cent in 2008.[37] The country is at risk of being left behind in this area unless the trend can be halted. The solution is to offer financial assistance to companies to innovate in the UK, to help them develop what is normally their most expensive and important competitive advantage. This also links into skills, as the motives for offshoring R&D are different to those of moving production. It requires less concern about costs, as the equipment used will consume most of the allotted money and instead, the greatest concern will be having access to a decent-sized pool of the highest-skilled workers available.

Innovation has been judged to be the touchstone for Britain's future as a manufacturing nation and according to Vince Cable, this means 'we need to earn our living in the world through high-tech, high-skills and innovation'.[38] According to Gerd Bender, innovation can also be 'incremental... reinforcing existing knowledge and competencies', or 'radical', which is a process of creative destruction.[39] A popular consensus has emerged that in the globally competitive market those most innovative industries in terms of product and process will stay ahead of the competition and be most successful.

However, the British government has become preoccupied with the formal form of innovation: research and development. This measure is the principal method by which innovation funding is allocated, but it does not accurately reflect the composition of British manufacturing or those who offshored production, and to adhere to this will mean many companies miss out on funding and support that could encourage them to bring their production home.

The Government is concerned with providing the most support for high-tech industries, which are those who spend over five per cent of their turnover on R&D. Thus David Cameron has publically proclaimed future support for: 'aerospace, pharmaceuticals, high-value manufacturing, hi-tech engineering, low carbon technology and all the knowledge-based businesses'.[40] However, only a minority of low-tech firms, those spending less than 0.9 per cent of turnover on R&D, have a formal R&D department and so for them, this means that the vast majority will miss out on the financial benefits the Government has planned to encourage innovation.[41] The number of manufacturers who agree with the Government's assertion that research is the most important source of competitive strength is small. In a survey, only two per cent of companies said this was true, as opposed to 30 per cent who said their production processes gave them the edge.[42] Perhaps most disturbing for the Government is that this survey included HT companies who, while spending lots on research, didn't see this as the prime source of their strength.

The model of encouraging formal innovation assumes that, within a business environment, increasing investment in R&D will lead to greater innovation and as a result, an advantage in the global economy. This concept has been increasingly criticised in recent years for being overly simplistic. Investment is also to be channelled into scientific academic centres, with the reasoning that, because Britain is a world leader in this form of research, funds will lead to greater numbers of discoveries that become 'spin-offs': viable companies that eventually expand the high-tech sector through venture capital backing. This is often called the linear model, which the Government also claims to reject while following its basic formula. David Willetts, Secretary of State for Universities, has claimed this 'sausage machine' thesis has been discarded, claiming 'the world does not work like this as often as you might think'.[43] The practical and immediate value of R&D is paramount for the Government and Vince Cable stated that, 'I support, of course, top class "blue skies" research, but there is no justification for taxpayers' money being used to support research which is neither commercially useful nor theoretically outstanding.'[44]

The crucial issue here is that the R&D focus fails to take into account intangible values such as the experience of the workforce which have huge effects on the ability of firms to recognise the need for innovations and implement them. In low-tech companies in particular, this informal innovation, where improvements emerge from all types of employees rather than specific researchers in dedicated departments, is key to their success. In other words, the UK's competitive advantage in these firms is not based on low costs, but on the intangible assets that make up their contribution to the 'knowledge economy'. This is why British low-tech production still has a surprisingly high comparative advantage, at around 92 per cent of the OECD average for basic goods in 2008.[45] In low-tech firms, employees' daily contact with the production processes means they gain a valuable insight into how to further improve these systems. The knowledge they gain from experience then manifests into practical advantages such as flexible production and increased innovation. The preoccupation with quantifiable indicators means that one of the greatest assets of production in the UK, the communal knowledge of the workforces, are being overlooked and their value understated, so that the worth of the low-tech sectors which rely on this most are, as a whole, undervalued.

Knowledge Transfer Partnerships

An important way that many British businesses use to overcome the expense of R&D is outsourcing this to British universities. This way, they gain the specialised, world-leading expertise needed without the permanent cost of maintaining a dedicated R&D department. The most successful model of businesses consulting academic institutions is the Knowledge Transfer Partnership scheme (KTPs), which was set up in 2003 and currently operates around 1,000 partnerships simultaneously.

In a KTP, a business pays for a high-calibre science or engineering graduate, under the auspices of a university, to tackle their specific R&D problem. Via the graduate, the company gains access to the physical resources of educational establishments and in return, these graduates gain the valuable industry-related experience many employers seek. There are numerous success stories that have been products of the KTP, winning all sorts of manufacturing awards and this feeling is certainly shared by the industrial community, with the number of businesses applying jumping from 177 in 2009 to 326 by September 2010, a record high for the scheme. In all, 6,000 graduate jobs have been created through KTPs and £4 billion generated in additional sales. However, the scheme had its funding cut in 2011, leading to a 25 per cent reduction in approved projects that year and it is expected that numbers will fall further through 2012. Given the scheme covers two-thirds of the cost of projects undertaken, the result could mean costs of KTP usage rise, effectively disincentivising innovation. This is particularly problematic as KTPs are especially valued by SMEs, who can take advantage of grants that mean the company only pays one-third of the KTP costs, on average an affordable £20,000. Larger businesses pay two-thirds of the costs. If this is subsidy is lost, the results will be dire, especially for manufacturers which are completely reliant on KTP services.

Better tax regime

The UK's taxation regime is perceived by many manufacturers to be a fairly significant barrier to growth and as was seen earlier, other countries offer special rates for offshorers. Britain offers no motivation for onshorers and until the UK attempts to reward companies for coming home, this will act against Britain. EEF's 2009 survey found that tax was cited as a bad factor when doing business in the UK by over 60 per cent of respondents, with over 20 per cent rating it as 'very bad'.[46] This feeling is reinforced by the World Economic Forum, which found tax rates to be the single most problematic factor cited by businesses and tax regulations to be the fourth greatest issue. Of course, while it is hardly surprising that businesses want to pay less tax, that Britain was ranked 94th for 'extent and effect of taxation' in this report shows that it is doing worse than many of its rivals.[47]

Corporation tax rates have been reduced by the current government and will be reduced to 23 per cent by 2014. This is better than nothing but given most offshorers already pay this, reducing it further will not encourage them to return actual production home (although it would act as a strong incentive for other companies to relocate to the UK which is certainly a good thing). Onshorers will be most concerned by the Government's failure to reform capital allowances and the 2012 Budget saw a reduction in capital allowances to claw back two-thirds of the cost of lowering the corporation tax rate. This was achieved by reducing the main recovery rate from 20 per cent to 18 per cent and furthermore, the annual investment allowance fell from £100,000 to £25,000. Previously, the Labour Government doubled the Annual Investment Allowance (AIA) from £50,000 to £100,000, with much positive feedback from industry. The latest move is highly contradictory to the UK's needs.

The AIA scheme has helped firms compete in the global market by offering tax relief on capital investment and this therefore gives firms an incentive to upgrade the equipment used in production. The example of America shows that many firms returning production to the UK will be looking to do this via advanced production processes which require less labour per unit of output. As Figure 2 (p.182) shows, this has already been occurring, but to continue and accelerate the trend, there will be much higher investments in capital. Wells Fargo, using the example of the US, found that this is even truer for foreign companies investing in American plants. Given that they could establish production anywhere, they chose the US for its high quality production processes and invested comparatively highly in capital goods. From 1997-2006, employees of subsidiaries were more productive than the US average manufacturing employee and became increasingly more so, widening the gap. From 2004-06, there was a three per cent gap.[48] Further strengthening the relationship between advanced production and worker skillset, employees of these more capitally intensive subsidiaries were also compensated 16 per cent higher in 2006 than the US manufacturing average, reflecting their higher qualifications and efficiency.[49] If the Government is to encourage onshoring, it must make it much easier for British companies to benefit from investing in British plants and machinery.

Reducing capital allowances penalises companies investing for the future and reduces the incentive to improve production processes at the time that developing countries will be increasingly competing with Britain in terms of advanced production.

The recent 2012 Budget was not completely negative for businesses though. The concept of the 'patent box' was introduced, which allows companies to pay lower tax rates on profits generated from IP held in the UK. This is most useful for pharmaceutical companies and was in part a reaction to the loss of Pfizer in February 2011 discussed below (see p.159). In principle, this means it is cheaper for companies to manufacture the drugs developed in the UK, and is therefore a welcome way to encourage holistic manufacturing. Its introduction has led GlaxoSmithKline to invest £500 million in a new UK plant and developments at two existing sites that will create 1,000 jobs in total. The company's CEO, Sir Andrew Witty said the patent box 'has transformed the way in which we view the UK as a location for new investments ensuring that the medicines of the future will not only be discovered, but can also continue to be made here in Britain.'[50] The patent box will not go unnoticed by offshorers and is certainly a step in the right direction.

Between patent boxes for knowledge-intensive firms, and better capital allowances for capital-intensive firms, these advances provide targeted but simultaneously general and far-reaching support that will go a long way towards incentivising onshoring.

Widen access to finance

Being able to raise a loan is still a difficulty for many British manufacturers and this makes the country considerably less attractive to businesses. The Global Competitiveness Report found that finance was the second most problematic issue cited with doing business in the UK. It also revealed that it is easier to access both loans and venture capital in China than in the UK.[51] Given that offshorers can often make deals with local Chinese officials to receive funding to set up their Chinese plants, or find outsourcing production can overcome their own inability to fund a new British plant, solving this problem will encourage onshoring. Additionally, onshoring would be a large expense for many manufacturers, who would have to invest large sums to set up new plants, bring machinery home and disrupt their production temporarily. The inability to access a decent-sized loan could be the deal breaker.

A lack of lending is one of the single most pressing issues facing the British economy. Small and medium-sized enterprises are the worst hit and without loans allowing them to expand or invest, economic growth will continue to be stunted. SMEs account for 99.9 per cent of all British enterprises, 60 per cent of private sector employment and 50 per cent of private sector turnover but with access to finance, they could expand further. At present, many British SMEs suffer from the 'Macmillan Gap', where funding from approximately £250,000 to £2 million is hard to come by as commercial lenders are uninterested in lending these sums. Worryingly, while the number of SMEs applying for loans has risen, so have the number of rejections. 35 per cent of SMEs sought finance in 2007 and 90 per cent of loan applications to banks were successful. In 2010, 42 per cent of SMEs sought finance but bank acceptance rates declined to 65 per cent.[52]

While the recession has weakened lending, perfectly sound businesses and entire sectors are being turned away for no reason. The only way to revive lending permanently is to go beyond just restructuring commercial banks. Britain needs the creation of a new state-backed investment bank. This would be able to raise cheap credit in the financial markets by using the UK's AAA credit rating and could pass this on to borrowers. It would lend to *any* SME rejected by commercial lenders but despite being economically viable. This bank would be most effective if it works with the commercial banks as middlemen but retains an expertise in assessing industrial businesses unseen in the UK for decades. By funding through quantitative easing measures already set to happen, the Bank could be established with billions of funds with no cost to the taxpayer. If such an institution is not created, Britain risks being unable to cope with the next recession and unable to attract the bigger offshorers home than need large loans for capital investment.

The Sheffield Forgemasters debacle of 2010 is now a classic example of how the Government does not understand the need for financial assistance for capital investment and growth. In March 2010, the company aimed to expand into the nuclear reactor parts market and secured a Labour government loan of £80 million to buy a 15,000-tonne press for this purpose. The loan was to be provided at a relatively low interest rate of 3.5 per cent, and made sense given the wider government policy of building new nuclear power stations and allowed a measure of import substitution. At the time, only one other firm in the world had the capacity for this, Japan Steel Works, and given the huge cost of shipping such large parts half-way across the world, this would have given Forgemasters a market not only in Britain, but Europe and even further afield.

However, the loan was then withdrawn by the Coalition Government in June 2010, as part of a £2 billion government loan cancellation that affected 12 projects in total, aiming to cut back on the budget deficit. This effectively scuppered Forgemasters' project. As such, the project had to be cancelled and the Board of Sheffield Forgemasters concluded: 'there is no easily available private sector alternative funding structure which is both economically viable for the Company and fair to existing shareholders'.[53] The only other option had been to finance through equity and give up control of the firm, which was not deemed acceptable so effectively, there was never a non-governmental way to access finance. The experience not only demonstrates the market failure in the provision of finance for longer term big investments, capital intensive investments and adjacent innovation but also highlights the role of government in providing signals to the private sector. The Labour Government took so long to announce its support of nuclear power that investors were unwilling to get involved when it finally did.

As a way to reduce the deficit, the withdrawal was very short-sighted: the government has only saved £80 million that would have been repaid anyway, given that this was a loan, not a grant. Additionally, with the nuclear industry beginning to revive after the 2011 Fukushima disaster, Forgemasters would have healthy demand and in time, paid more in tax revenues to the State, as would its increased workforce. Moreover, any new British nuclear power stations will need to import their components from Japan, worsening the trade deficit as well. The cancellation has therefore come at the cost of long-term benefit to Forgemasters, and Britain as a whole. In October 2011, the company was given a loan of £36 million via the Regional Growth Fund, but for general capital investment and, as Ed Miliband described it, this was 'too little, too late'.[54] The whole issue could have avoided through the use of an industry bank which could take these sort of decisions based on economic rather than political judgements.

Without increased access to finance, the onshoring trend is unlikely to develop much further. Companies will not base themselves in locations that constrain rather than enable growth. It is the Government's responsibility to overcome the failure of our commercial banks and to fund onshoring. Not only will this enable companies, but it will also prove their commitment to support the trend.

Export finance

As already stated, one of the key reasons large firms choose to base themselves in the UK is its position as the 'gateway' to Europe and wider markets. However, this belies the fact that many firms, particularly SMEs, find exporting very difficult financially, and therefore have less incentive to onshore. This difficulty occurs for a number of reasons. Key among these is the lack of a government-backed trade finance regime for companies. In many countries it is a legal requirement to have export trade credit insurance and even in those where it is not, the state frequently provides the means for this or comparable levels of support. In Britain, there are many examples of how the private market has failed to adequately provide this but the government has still not stepped in.[55] Countries across Europe and the developed world have this, leaving British companies isolated. Moreover, private insurers are increasingly refusing to insure businesses simply on the basis that they work in a particular sector that they refuse to get involved in. In the end, the British manufacturer then loses out on a contract to a foreign company whose goods might be more expensive or of lower quality and that would have the same problem on the private market, but can access state facilities instead. In 2010, *The Economist* analysed the problem in depth:

> The Export Credits Guarantee Department (ECGD)... sees its role less as an active promoter of British exports than as a last-ditch line of defence when the market has no answer (it is also charged with making a profit). In 1991 it stopped providing short-term export-credit guarantees, spinning off the business to what is now the privately owned Atradius. ECGD reserves its firepower mainly for long-term contracts that are too big and risky for the private sector—70% of its portfolio now consists of aerospace exports.[56]

During the recession, the EU acknowledged the difficulties manufacturers were having in securing short-term contract export credit, and relaxed its rules accordingly. France, Germany and the Netherlands all applied to take advantage of the new freedoms but the

UK did not, arguing that the private credit insurance market would recover. This has not been the case.

The problem here is twofold. Firstly, this clearly restricts the potential of firms to expand through foreign orders with all the added benefits this would have for the UK economy. Orders from abroad have to be turned down even if the company is successful and the potential failure to deliver is negligible. In the long-term, this is very damaging, as British firms are slowly squeezed out of international markets by foreign competitors who have no such handicap. Secondly, a potentially more dangerous issue is that companies who rely on exports for their survival have found that as costs for credit insurance on the private market has risen, they are forced to cut back production, threatening the business. The government needs to implement a trade credit scheme, not so much to favour British manufacturing but just to ensure it is not unfairly disadvantaged as is currently the case. This will remove one more offshoring 'push' factor and replace it with an onshoring 'pull' one.

Create a holistic environment

Supply chain resilience, a key reason for onshoring, can be promoted through the creation of an industrial policy that values all the types of companies in a supply chain, not just the end one. This also goes for the activities of the companies. It has been received wisdom for some time that a company can keep its innovation centres at home while offshoring the actual production processes to dedicated plants elsewhere. For Britain as a whole, this means reinventing the country as a global hub for R&D. Such an idea encourages offshoring on practical level.

This has been the basis for Coalition policy for some time with the idea stemming from the report James Dyson wrote for the Conservatives back in 2010 prior to the election.[57] This does indeed seem to be Dyson's own business model, as his research is carried out in Britain, but actual production was moved from Wiltshire to the Far East in 2002, at a loss of 800 jobs.[58] Despite wanting to encourage more British manufacturing, at the time Dyson tried to justify this move by saying: 'I don't think I can (see an alternative), it's been an agonising decision and very much a change of mind. Increasingly in the past two to three years our suppliers are Far East based and not over here, and our markets are there too.'[59] However, evidence would suggest this is a rare success and not seen as desirable by many companies. Respondents to a 2007 survey said they expected to continue with their production in Britain in five years' time. Other divisions were not much safer and UK-based research was only predicted by 72 per cent of 1,000 respondents to remain here.[60] Keeping manufacturing in the UK is an 'all or nothing' business. Most SMEs do not have the luxury of splitting the company like this anyway but when operations in larger companies are split, in time, the research departments often end up emigrating as well. A more recent EEF survey examined the consequences of offshoring production, effectively testing Dyson's argument. It found that in companies producing their goods solely in the UK, over 80 per cent also had their innovation centres entirely in Britain. For those producing in both developed and emerging countries, just under 50 per cent retained just British innovation outlets and around 40 per cent had innovation centres in other developing and emerging economies.[61] Dyson's concept does not hold up to scrutiny and aggravates the offshoring issue.

Pfizer was a surprise example of this emigrating trend when it announced in February 2011 that it was shutting down its R&D facility in Kent, at a cost of 1,600 jobs with 900 staff retained.[62] Pfizer had closed down its manufacturing facilities on the site in 2007, losing 420 employees. This calls into the question the ability to separate manufacturing and its research: the government should not be satisfied with the shaky commitments made to retain the intellectual operations in the UK and should seek to make Britain suitable for the whole business process. After all, the press release given out by Pfizer after the 2007 cutback to the production process stressed:

> Pfizer remains committed to its research and development programme at Sandwich which, as announced earlier this year, is one of Pfizer's four key global research and development sites.[63]

Pfizer's words and actions were very different.

Rolls-Royce displays the optimal solution for many companies: it currently bases research and production on the same campus in Derby. The reasoning is that a close relationship between designers, engineers and factory-floor workers will create stronger ties and understanding of their respective roles. In turn, it is expected that this will create more of the paradigm shifts in innovation that Rolls values rather than just incremental advances. This also allows the traditional three-step process of design, materials selection and manufacture to be merged into one holistic process. Given a product could fail at any one of these stages, this means designers spend less time on ideas that are impractical, and each group can refine the eventual product.

Foster inter-British industry relationships

To encourage the onshoring of existing outsourced contracts, the government should reinstate the service that used to be provided by the now defunct Innovation Advisory Service (IAS). This organisation acted effectively as a database but, unlike conventional ones that just list producers, was designed for businesses with specific problems to advertise for companies to provide solutions. The IAS would then contact other companies detailing the issue and effectively run a competition for regional firms to fight to provide a solution. This open innovation model worked very well and created many inter-British industry partnerships and was used by very large firms such as BAE Systems when in-house R&D departments, or existing external suppliers, could not solve their problems.

On the whole, communications between firms not already linked by supply chains is often poor, so this type of database is a necessary and effective way to combat the problem. In a modern incarnation, a 'problem/solution database' would need to perform both roles and provide two-way communications. Ideally, suppliers would be able to approach potential customers and offer solutions in terms of product or process that the customer was not even aware of. The results of this would be manifold, from import substitutions when it is realised required components are already created in this country, to allowing market expansion. While there are trade organisations that provide this service to a degree, these usually restrict relations to specific sectors by default, narrowing the potential for taking advantage of external knowledge. Also, whilst there are frequent trade conferences in Britain which connect companies, these are temporary and do not deliver a permanent solution or impose a framework. A dual database of suppliers and problems would be a valuable addition to an existing patchy network of British companies. This is already being promoted within certain sector-based trade organisations to avoid offshoring. The Society of Motor Manufacturers and Traders has been working to bring together original equipment manufacturers and suppliers to match them up. In addition, and with the assistance of the Manufacturing Advisory Service, it advises and helps domestic component suppliers to actually win contracts. While progress has been made in the automotive sector, there are many other industries in real need of a similar service to avoid the necessity of offshoring.

Manage and restrain cluster policy

The Government is focused on creating more industrial clusters, having seen the success of those such as the Cambridge technology cluster and wanting to replicate this in other locations and sectors. While this looks good on paper, it is not necessarily the way to encourage onshoring as done badly, it adds no additional value to supply chain resilience.

The Department for Business, Innovation and Skills defines clusters as 'geographic concentrations of interconnected companies, specialised suppliers, service providers, firms in related industries, and associated institutions... they compete but also co-operate'.[64] Clusters are an ideal way to encourage onshoring and prevent offshoring occurring in the first place. Their existence is perfect for spin-out companies who lack the financial and technical resources to independently commercialise their ideas. The cluster provides an agglomeration effect and a 'sandpit' where ideas can be generated and developed. This is mainly required at the alpha, beta and gamma stages of production so by teaming up with other similar companies, often in the shade of successful universities, a hospitable environment of shared tangible and intangible resources is created for companies and skills to grow in. The testing and prototyping stage of development is one of the biggest costs to businesses given that the facilities for this have to be created, if they are not already available, so David Willetts was right to support a pooling of these:

> It makes sense for government to back shared facilities – research platforms if you like – which private companies could not develop on their own... this is how publically backed R&D boosts economic performance.[65]

This prevents a forced emigration to countries where all the funds would otherwise be available and such an approach should continue to ensure UK-nurtured firms develop in Britain and the eventual benefits of fully-fledged companies paying tax and providing employment likewise remain here.

Clusters are nothing new and have existed UK-wide for some time such as in Dundee which is now a computer games development cluster. The Dundee cluster has developed well, producing 10 per cent of the digital entertainment output of Britain and an annual turnover of £100 million.[66] It is therefore no surprise that David Cameron aims to create more and said: 'let's make Humberside lead the world in carbon capture and storage. Let's make Bristol a centre for marine energy parks'.[67] However, creating clusters from scratch is not feasible and would incur large costs for potentially little returns. The simple reason is that if there was a need for a cluster, it would already exist or its beginnings would be evident. Moreover, the purpose of such clusters is seen by the Government in terms of allowing a pooling of research resources, and therefore for high-tech companies rather than a place for their suppliers or less R&D intensive sectors as well: they appear to be modelled on science parks. The Government has made no mention of helping non-high-tech manufacturers and suppliers develop within these clusters to reinforce themselves and their customers. This would unlock the practical advantages of geographical proximity discussed earlier (see p.102) and reinforce the productivity off all levels of the supply chain, and boosting the economy both regionally and nationally.

Another key issue is that there has been a continual drive towards creating unnecessary clusters through the regionalisation of cluster policy. The now defunct Regional Development Agencies all tried to create their own lucrative biotech and electronic clusters within their spheres of influence but by pulling sectors in all geographical directions at once, the overall advantage that clusters could provide in terms of concentrating knowledge was diluted, so the value of individual clusters is poorer. Clusters are just that, concentrations of certain sectors, so the aim of creating more than is necessary is self-defeating. The Coalition's successor organisations, the newly launched Local Enterprise Partnerships (LEPs), work at an even more localised level which may lead to an even more blinkered scramble for clusters. The aim is to cut red tape and ensure financial aid is directed to wherever it is needed the most, effectively performing part of the role of the RDAs, while the Government has stated that these will retain control of 'inward investment, sector leadership, business support, innovation and access to finance'.[68] The best outcome would be one in which the Government moves away from focusing on creating new clusters to using their retention of national strategy to ensure existing clusters are sustained and incentivised. An alternative would be to allow LEPs to bid for funding for already developing clusters in their area as they should have more awareness as to where money is needed the most to bolster the supply chain. The third round of funding of the £1 billion Regional Growth Fund could be best used to this purpose, providing money to areas with the best prospects and ensuring through competition that the same sector cannot be spread too thinly. This is the best way to offer a level of supply chain resilience that will attract offshorers home.

Conclusion

The migration of industries and companies is as natural a phenomenon as the migration of people and, like people, they will move to wherever they believe the conditions for their existence are optimal. For many British manufacturers, China and other emerging economies have offered better conditions over the last decade, so they relocated accordingly. Naturally, as the advantages are dwindling, some are returning of their own accord and many others could be persuaded to come home.

The experience of these offshoring and onshoring companies teaches us that economic climates are constantly in flux, and what might have seemed inevitable or permanent ten years ago is certainly not so. Likewise, if the British government fails to create a positive economic environment for British manufacturing and the optimal conditions for economic growth, then there is no reason to suppose the onshoring trend will last long, and it could be a brief reprieve from a more sustained migration abroad.

Britain is currently at a critical juncture in terms of the size of its industrial economy, and academics and experts are undecided about the extent to which the UK has lost its industry. Some argue that the UK is at risk of losing its critical mass of baseline industries as increasing numbers offshore production or outsource it beyond the UK, while others suggest we have already lost this. Once industry dwindles beyond this point, it is effectively lost permanently and the entire web of interlinked manufacturers unravels. Onshoring works in opposition to this force and as companies return production home there is an increasing need for domestic suppliers and skilled workers, creating the extra demand in the UK that can encourage further British manufacturers to onshore industries. This is especially true for the companies that left the UK in the first place to locate production nearer to their customers.

A thorough analysis of the potential for American onshoring found that the 'biggest constraint on jobs is business uncertainty'.[1] While it might or might not be the number one factor in the UK, without certainty that the skills base will increase, the financial and regulatory environments will improve, and explicit support for manufacturers will emerge, British businesses will have no reason to bring production home. The Government must commit to them to ensure businesses commit to the UK. Without this, some companies, if they are not certain that future conditions favour repatriation, would continue to minimise their risk by offshoring or outsourcing production rather than make the investments onshoring needs.

If Britain is to generate further onshoring and retain manufacturers, it needs to stay ahead of the international competition and help its industries evolve in the various ways needed in order for them to survive and thrive.

Notes

Introduction

[1] US-China Business Council website, 'China's world trade statistics', table 8

[2] EEF & BDO, *Manufacturing Advantage: How manufacturers are focusing strategically in an uncertain world*, November 2009, pp. 6-7

[3] UNCTAD statistics, 'Values and shares of merchandise exports and imports', annual, 1948-2010 & ONS, GDP Deflators at market prices, and money

[4] Cable, V., speech, *Speech to the Liberal Democrat Conference*, 22 September 2010

[5] *Manufacturing Advantage*, p. 8

[6] *Manufacturing Advantage*, p. 12

[7] OECD statistical database, *Glossary of Statistical Terms*, Outsourcing

[8] EEF/BDO, *Global Challenge Survey*, February 2008, p. 8

[9] McKinsey, *Exploding the myths of offshoring* pp. 1-5

[10] Zeng, A., & Rossetti, C., 'Developing a framework for evaluating the logistics costs in global sourcing processes', *International Journal of Physics Distribution and Logistics Management*, Vol. 33, 9, pp. 786 - 792

[11] *Manufacturing Advantage*, p. 8

[12] *Manufacturing Advantage*, p. 11

[13] *Global Challenge Survey*, p. 3

[14] *Manufacturing Advantage*, p. 11

[15] Platts, K., & Song, N., *The true costs of overseas sourcing*, paper given at the POMS 20th Annual Conference, May 2009, p. 20

[16] Song, N., Platts, K., & Bance, D., 'Total acquisition cost of overseas outsourcing/sourcing: a framework and a case study', *Journal of Manufacturing Technology Management*, Vol. 18, 7, 2007 pp. 858 – 875

---

[17] For a concise table of these, see *The true costs of overseas sourcing*, p. 6, table 1
[18] 'Total acquisition cost of overseas outsourcing/sourcing', p. 871
[19] 'Total acquisition cost of overseas outsourcing/sourcing', p. 872
[20] *The true costs of overseas sourcing*, p. 6
[21] American Institute for Economic Research website, 'The Return of U.S. Manufacturing', 27 July 2011
[22] ONS, *The Pink Book 2011*, February 2012, Table 1.1
[23] BBC news, 'Tata Steel to cut 1,500 jobs in Scunthorpe', 20 May 2011 accessed here: http://www.bbc.co.uk/news/business-13469088
[24] ONS, *Annual Survey of Hours and Earnings, 2011 Provisional results*, Table 16.1a
[25] Cook Associates website, 'Rising Labor Costs and Quality Concerns'
[26] *FT Alphaville* 'The return of the US manufacturer', 4 April 2012
[27] OECD statistical database, *STAN R&D expenditures by industry*, UK, national currency
[28] Hirsch-Kreinsen, H., Jacobson, D., & Robertson, P., (eds.), *'"Low-tech" Industries: Innovativeness and Development Perspectives. A summary of a European Research Project'*, December 2005, pp. 22-3
[29] Amiti, M., & Wei, S.-J., *Fear of Service Outsourcing: Is it Justified?*, IMF Working Paper WP/04/186, October 2004, p. 6
[30] *Fear of Service Outsourcing*, p. 20
[31] World Trade Organisation, *International Trade Statistics 2011*, tables III.14
[32] *Manufacturing Advantage*, p. 4
[33] UNCTAD statistics, 'Values and shares of merchandise exports and imports', annual, 1948-2010
[34] 'Values and shares of merchandise exports and imports'

---

[35] World Trade Organisation, *International Trade Statistics 2011,* table I.10
[36] *Global Challenge Survey,* p. 3
[37] *Made in America, Again,* p. 12
[38] Boston Consulting Group, *Made in America, Again: Why Manufacturing Will Return to the U.S.,* August 2011, p. 12
[39] Boston Consulting Group, *U.S. Manufacturing Nears the Tipping Point: Which Industries, Why, and How Much?,* March 2012, p. 11
[40] *Global Challenge Survey,* pp. 3-5

1 What drives onshoring?

[1] *The Sunday Times,* 'Made in Britain: How manufacturing is returning to the UK', 3 January 2010
[2] *Manufacturing Advantage,* p. 14
[3] Platts, K., & Song, N., *The true costs of overseas sourcing,* paper given at the POMS 20th Annual Conference, May 2009, p. 18
[4] *The true costs of overseas sourcing,* p. 18
[5] *The Sunday Times,* 'Made in Britain'
[6] Song, N., Platts, K., & Bance, D., 'Total acquisition cost of overseas outsourcing/sourcing: a framework and a case study', *Journal of Manufacturing Technology Management,* Vol. 18, 7, 2007 p. 858
[7] Bureau of Labour Statistics, *International Comparisons of Hourly Compensation Costs in Manufacturing,* Table 1.2 Hourly compensation costs in manufacturing, U.S. dollars, 1996-2010
[8] *International Comparisons of Hourly Compensation Costs,* Table 1.2 & *U.S. Manufacturing Nears the Tipping Point* p. 3
[9] *Wall Street Journal,* 'China's wage hike ripples across Asia', 13 March 2012
[10] *Made in America, Again,* p. 7
[11] *U.S. Manufacturing Nears the Tipping Point,* p. 6
[12] 'The Return of U.S. Manufacturing'

[13] ONS, *2011 Annual Survey of Hours and Earnings (based on SOC 2010)*, April 2011, Table 3.5a Median full-time gross hourly earnings by region
[14] 'The Return of U.S. Manufacturing'
[15] *New York Times*, 'Foxconn Increases Size of Raise in Chinese Factories', 6 June 2010
[16] *The Register*, 'Embattled Foxconn raises wage slaves' salaries, 17 February 2012
[17] http://chinaautoweb.com/2010/06/honda-lost-yuan-3-billion-in-sales-on-strike/
[18] *BBC News*, 'Apple and Foxconn plan raises bar for Chinese factories', 4 April 2012
[19] ONS, *Industries intermediate consumption in 2009*, The 'Combined Use' matrix, November 2011, Table 2 Int Con 2009, compensation of employees divided by total output at basic prices.
[20] *The Guardian*, 'Why the future is made in Britain', 27 April 2008
[21] *Manufacturing Advantage*, p. 14
[22] *The Economist*, 'The end of cheap China', 10 March 2012
[23] 'The Return of U.S. Manufacturing'
[24] China Daily news, 'Wage hike to benefit migrant laborers', 3 March 2011, accessed here:
http://www.chinadaily.com.cn/china/2011-03/03/content_12106767.htm
[25] *The Economist*, 'The end of cheap China', 10 March 2012
[26] *The Economist*, 'The boomerang effect', 21 April 2012
[27] BBC documentary, *The town that took on China*, episode 1, broadcast 8 May 2012
[28] *The town that took on China*, episode 1
[29] *The New York Times*, 'How the U.S. Lost Out on iPhone Work', 21 January 2012
[30] World Bank
[31] BBC documentary, *The town that took on China*, episode 2, broadcast 15 May 2012

[32] Economist Intelligence Unit, *Gearing for growth: Future drivers of corporate productivity*, March 2011, p. 28
[33] *Gearing for growth: Future drivers of corporate productivity*, pp. 29-30
[34] Bureau of Labor Statistics, U.S. Department of Labor
[35] House of Commons Library, *International comparisons of manufacturing output*, January 2012 & ONS dataset, 'All in employment by industry sector, EMP13
[36] Centre for Research on Socio-Cultural Change, *Apple Business Model: Financialization across the Pacific*, April 2012, p. 17
[37] *Apple Business Model*, p. 17
[38] *Apple Business Model*, p. 18
[39] *The town that took on China*, episode 2
[40] *The town that took on China*, episode 2
[41] Cooper, R., Gong, G. & Ping, Y., *Costly Labor Adjustment: Effects of China's Employment Regulations*, National Bureau of Economic Research, Working Paper 17948, March 2012, p. 5
[42] *Costly Labor Adjustment*, p. 5
[43] http://moneymorning.com/2008/01/03/new-labor-laws-and-strengthening-yuan-could-put-the-squeeze-on-chinese-exports/
[44] 'Total acquisition cost of overseas outsourcing/sourcing', p. 861
[45] *Telegraph*, 'David Cameron should be focusing on intellectual property not human rights', 9 November 2010
[46] 'Total acquisition cost of overseas outsourcing/sourcing', p. 861
[47] Capaccio, T., 'China Top Source of Counterfeit U.S. Military Electronics', 22 May 2012: http://www.bloomberg.com/news/2012-05-21/china-top-source-of-counterfeit-u-s-military-electronics.html
[48] *Manufacturing Advantage*, pp. 7-17
[49] Rilla, N., & Squicciarini, M., 'R&D (Re)location and Offshore Outsourcing: A Management Perspective', *International Journal of Management Reviews*, Vol. 13, 2011, p. 395

[50] Priest, E., 'The Future of Music and Film Piracy in China', *Berkeley Technology Law Journal*, Vol 21, 2006

[51] Thomson Reuters, *Spot Prices for Crude Oil and Petroleum Products*, April 2012, RBRTE

[52] *U.S. Manufacturing Nears the Tipping Point*, p. 7

[53] *Daily Telegraph*, 'Manufacturing returns to Britain', 12 July 2010

[54] 'Total acquisition cost of overseas outsourcing/sourcing', p. 870

[55] *The Economist*, 'The end of cheap China', 10 March 2012

[56] *The true costs of overseas sourcing*, p. 16

[57] *The true costs of overseas sourcing*, p.17

[58] *China Daily*, 'Industry faces rising power cost', 31st May 2011

[59] http://news.xinhuanet.com/english2010/china/2011-01/17/c_13693802.htm

[60] 'Industry faces rising power cost'

[61] *Made in America, Again*, p. 10, conversion from square feet and dollar costs

[62] Valuation Office Agency, *Property Market Report* 2011, January 2011, p. 29, from value in hectares

[63] http://www.indexmundi.com/commodities/?commodity=iron-ore&months=60

[64] http://www.indexmundi.com/commodities/?commodity=industrial-inputs-price-index&months=120

[65] *The Guardian*, 'Why the future is made in Britain', 27 April 2008

[66] www.xe.com, Currency historical charts, GBP per 1 CNY

[67] *U.S. Manufacturing Nears the Tipping Point*, p. 7

[68] *Manufacturing Advantage*, p. 15

[69] www.xe.com, Currency historical charts, GBP per 1 U.S.D

[70] http://www.tradereform.org/2011/07/cpa-white-paper-how-chinas-vat-massively-subsidizes-exports/

[71] CRESC, *Apple Business Model: Financialization across the Pacific,* April 2012, pp. 15-16
[72] M. Brooks, 'An eye on the prize', *New Statesman,* 16 August 2010, pp. 36-7
[73] Pecht, M. & Zuga, L., 'China as Hegemon of the Global Electronics Industry: How It Got That Way and Why It Won't Change', *IEEE Transactions on Components and Packaging Technologies,* 2008
[74] Zhao, Z., Huang, X., Ye, D., Gentle, P., 'China's Industrial Policy in Relation to Electronics Manufacturing', *China & World Economy* 15, (3), 2007, pp. 33-51
[75] 'Total acquisition cost of overseas outsourcing/sourcing', p. 858
[76] *Global Challenge Survey,* p. 9
[77] 'Total acquisition cost of overseas outsourcing/sourcing', p. 861
[78] Platts, K., & Song, N., *The true costs of overseas sourcing,* paper given at the POMS 20th Annual Conference, May 2009, p. 13
[79] 'Total acquisition cost of overseas outsourcing/sourcing', p. 861
[80] *Jakarta Post,* 'Amid many challenges, Vietnam's star continues to rise, 30 April 2012
[81] BBC News website, 'Which is the world's biggest employer?', 20 March 2012

2 The United States as a case study

[1] *Financial Times,* 'Business returns to U.S. as Asia loses edge', January 17 2012 & *FT Alphaville* 'The return of the US manufacturer', 4 April 2012. UK percentage calculated from ONS, *Pink Book 2011,* exports of all semi & finished goods, food, beverages and tobacco and basic materials divided by total exports of all goods and all services.
[2] *U.S. Manufacturing and the Economic Outlook*

---

[3] Pianalto, S., speech at the University of Toledo, Ohio, *U.S. Manufacturing and the Economic Outlook*, 20 October 2011
[4] Boston Consulting Group, *U.S. Manufacturing Nears the Tipping Point: Which Industries, Why, and How Much?*, March 2012, p. 6
[5] 'The Return of U.S. Manufacturing'
[6] 'The return of the US manufacturer'
[7] *Made in America, Again*, p. 3
[8] 'The return of the US manufacturer'
[9] Obama, B., speech, 'President Obama's State of the Union Address', 25 January 2012
[10] Cook Associates website, 'Rising Labor Costs and Quality Concerns Have Companies Re-evaluating Overseas Strategies', December 2011
[11] 'Rising Labor Costs and Quality Concerns'
[12]
http://www.appliancemagazine.com/news.php?article=1544614
[13] http://www.reshorenow.org/
[14]
http://www.supplychainquarterly.com/topics/Manufacturing/201104reshoring/
[15] Mars website, 'Mars breaks ground on first phase of new facility in Topeka KS', August 2011
[16] American Institute for Economic Research website, 'The Return of U.S. Manufacturing', 27 July 2011
[17] *Business Week*, 'Ford adds 12,000 Hourly Jobs in U.S. Plants under UAW Accord', 4 October 2011
[18] *Made in America, Again*, p. 12
[19] 'The boomerang effect'
[20] *U.S. Manufacturing Nears the Tipping* Point, pp. 3, 8 &12
[21] Based on an unemployment rate of 2.6 million (8.6 per cent) in March 2012. Source: ONS, *Summary of National Labour Force Survey Data*, AO2
[22] *The New York Times*, 'How the U.S. Lost Out on iPhone Work', 21st January 2012
[23] 'How the U.S. Lost Out on iPhone Work'

[24] 'How the U.S. Lost Out on iPhone Work'

[25] Xing, Y., & Detert, N., *How the iPhone Widens the United States Trade Deficit with the People's Republic of China*, December 2010, revised May 2011, p. 5

[26] *The Guardian*, 'Apple: why doesn't it employ more US workers?', 23 April 2012

[27] 'The boomerang effect'

[28] *How the iPhone Widens the United States Trade Deficit*, p. 6

[29] 'Apple: why doesn't it employ more US workers?'

[30] Pianalto, S., speech at the University of Toledo, Ohio, *U.S. Manufacturing and the Economic Outlook*, 20 October 2011

[31] Wells Fargo Economics Group, *Insourcing: Manufacturing – A Viable Solution in a Global Economy?*, February 2012, p. 1

[32] Rolls Royce press release, 9 March 2012. Accessed here: http://www.rolls-royce.com/northamerica/na/news/2012/120309_president_obama_visits.jsp

[33] *Financial Times*, 'Business returns to U.S. as Asia loses edge', January 17 2012

[34] BLS, Employment by industry, occupation, and percent distribution, 2010 and projected 2020, data series 31-330 (Manufacturing), March 2012

[35] 'Business returns to U.S. as Asia loses edge'

[36] *Insourcing: Manufacturing – A Viable Solution in a Global Economy?*, p. 4

[37] *U.S. Manufacturing and the Economic Outlook*

[38] 'How the U.S. Lost Out on iPhone Work'

[39] ONS, *UK Business: Activity, Size and Location*, 2011, Table B2.1 & United States Census Bureau, *2009 Statistics of U.S. Businesses Data*, November 2011, NAICS sectors, large employment size

---

3 What sorts of companies are most likely to onshore back to Britain?

[1] Cook Associates website, 'Rising Labor Costs and Quality Concerns'
[2] *Manufacturing Advantage,* pp. 16
[3] *The Economist,* 'Forging ahead' 21st April 2012
[4] EEF/BDO, *Global Challenge Survey,* p. 6
[5] *Global Challenge Survey,* p. 7
[6] STAN Database for Structural Analysis, Export of goods at current prices
[7] *Manufacturing Advantage,* pp. 13-14
[8] *The town that took on China,* episode 1
[9] EEF, *High value – How UK manufacturing has changed,* November 2007, p. 11
[10] *The Guardian,* 'Toyota profit slides on Japan earthquake disruption', 11 May 2011
[11] Rilla, N., & Squicciarini, M., 'R&D (Re)location and Offshore Outsourcing: A Management Perspective', *International Journal of Management Reviews,* Vol. 13, 2011, p. 400
[12] *Made in America, Again,* p. 13
[13] *The Economist,* 'Burgeoning bourgeoisie', 12 February 2009
[14] *The Sunday Times,* 'Made in Britain: How manufacturing is returning to the UK', 3 January 2010
[15] 'Made in Britain'
[16] *Global Challenge Survey,* p. 6
[17] *The Economist,* special report, 'A third industrial revolution', April 21 2012
[18] 'Forging ahead'
[19] 'A third industrial revolution'
[20] EEF, *High value,* p. 12
[21] 'Why the future is made in Britain'
[22] *The Guardian,* 'Why the future is made in Britain', 27 April 2008
[23] *Global Challenge Survey,* p. 7

[24] Hirsch-Kreinsen, Jacobson & Robertson (ed.), *PILOT Project Summary*, p. 16
[25] Coutts, K., & Rowthorn, R., *Prospects for the UK Balance of Payments*, London, Civitas, March 2010, p. 16
[26] *Prospects for the UK Balance of Payments*, p. 16
[27] *The Economist*, 'Solid print', 21 April 2012
[28] 'Solid print'

4 How should Britain encourage onshoring?

[1] Markit/CIPS News release, 'Markit/CIPS UK Manufacturing PMI', 2nd April 2012, p. 1
[2] 'Markit/CIPS UK Manufacturing PMI', p.2
[3] *Prospects for the UK Balance of Payments*, p. 12
[4] World Economic Forum, *Global Competitiveness Report 2011 – 2012*, Geneva, September 2011, p. 15
[5] *Global Competitiveness Report 2011 – 2012*, p. 360
[6] *Global Challenge Survey*, p. 7
[7] *The Telegraph*, 'Debt crisis: as it happened', 15 March 2012
[8] Boston Consulting Group, *U.S. Manufacturing Nears the Tipping Point: Which Industries, Why, and How Much?*, March 2012, p. 4
[9] *Made in America, Again*, p. 13
[10] OECD, STAN indicators ed. 2009, Value added shares relative to manufacturing
[11] Cable, *'Science, Research and Innovation'* HM Treasury, *'Budget 2011'* http://cdn.hm-treasury.gov.uk/2011budget_complete.pdf
[12] Full results in Hirsch-Kreinsen, H., Jacobson, D., & Robertson, P., (eds.), *'"Low-tech" Industries: Innovativeness and Development Perspectives. A summary of a European Research Project'*, December 2005
[13] Hirsch-Kreinsen, 'Low-Technology,' p. 5
[14] EEF, *High value*, p. 12
[15] OECD, 'Education at a glance 2005', Paris, 2005 OECD, *Country Background Report: Adult Basic Skills and Formative Assessment Practices*, 2005, p.3

[16] *Country Background Report: Adult Basic Skills and Formative Assessment Practices*
[17] Peacock L., 'Engineering companies give up on hiring school leavers', *The Daily Telegraph,* 9 September 2010
[18] OECD, *PISA 2009,* PISA country profiles , all students mean
[19] OECD, *PISA 2006 results,* figures 2.11c, 6.8b & 6.20b
[20] 'EU 2020 Strategy Executive Summary' (March 2010) p. 3
[21] Cable, V., speech, *Higher Education* 15 July 2010
[22] 'Engineering companies give up on hiring school leavers'
[23] Engineering and Machinery Alliance, *UK manufacturing in transition* (2003), p. 13
[24] *The Independent,* '24,000 chase just 221 apprenticeships', 16 August 2010
[25] The Data Service, *Apprenticeship Starts,* tables S6.1 & S6.2
[26] The Data Service, *Statistical First Release,* March 2012, Table 3.1
[27] WorldPublicOpinion.org, *Public Opinion on the Global Economic Crisis,* (21 July 2009), p.7
[28] 'The Return of U.S. Manufacturing'
[29] Merlin-Jones, D., 'Time for turning? Why the Conservatives need to rethink their industrial policy (if they have one)', *Civitas Review,* Vol. 7, 1, January 2010, p. 3
[30] *Daily Mail,* 'British car industry wins 'big vote' of confidence after Nissan announces plans to invest £127m and create more than 1,000 new jobs at UK plant', 10 April 2012
[31] Cameron, D., speech, *'Speech to the CBI',* 25 October 2010
[32] *BBC News,* 'U.S. Congress committee approves China sanctions bill', 24 September 2010
[33] For a longer discussion of the smelter closure, see Merlin-Jones, D., *The closure of the Lynemouth aluminium smelter: an analysis,* April 2012
[34] *Financial Times,* 'Doubt cast over power plant's future', 23 April 2010. Accessed: http://www.ft.com/cms/s/0/d7529c58-4e39-11df-b48d-00144feab49a.html#axzz1qDYSkOsY

[35] Rio Tinto press release, 'Rio Tinto Alcan announced intention to close Lynemouth aluminium smelter, 16 November 2011. Accessed: http://www.riotinto.com/media/5157_21255.asp

[36] http://www.advfn.com/nyse/StockNews.asp?stocknews=RTP&article=44499894

[37] OECD, STAN indicators ed. 2009, *Export market share relative to the world*

[38] Cable, V., speech, *'Science, Research and Innovation'*, Queen Mary University of London, 8 September 2010

[39] Bender, G., 'Innovation in Low-tech', *Arbeitspapier*, 6, September 2004, p. 9

[40] Cameron, D., speech, *'Transforming the British economy'*, 28 May 2010

[41] 'Innovation in Low-tech', p. 9

[42] *High value – How UK manufacturing has changed*, p. 9

[43] Willetts, D., speech, *'Science, Innovation and the Economy'*, The Royal Institution, London, 9 June 2010

[44] *'Science, Research and Innovation'*

[45] OECD, STAN indicators ed. 2009, *Index of revealed comparative advantage.* By way of comparison, in 2009, Korea has a high specialisation in high-tech manufacturing at around 150 per cent of the OECD average and Italy has a high low-tech specialisation at 171 per cent.

[46] *Manufacturing Advantage*, p. 17

[47] *Global Competitiveness Report 2011 – 2012*, pp. 360-1

[48] *Insourcing: Manufacturing – A Viable Solution in a Global Economy?*, p. 3

[49] *Insourcing: Manufacturing – A Viable Solution in a Global Economy?*, p. 3

[50] *Management Today*, 'Shot in the arm to UK manufacturing as GlaxoSmithKline invests £500m', 22 March 2012

[51] *Global Competitiveness Report 2011 – 2012*, pp. 149 & 360-1

[52] Office for National Statistics, *Access to Finance 2007 and 2010*, 28 October 2010, p. 1

[53] Sheffield Forgemasters website, 'Joint Funding Statement by Sheffield Forgemasters and HM Government', July 2010
[54] BBC News, 'Sheffield Forgemasters gets up to £36m from government', 31 October 2011
[55] British Chambers of Commerce, *Exporting Britain,* June 2009
[56] *The Economist,* 'Trading out of trouble', 18 February 2010
[57] Dyson, J., *Ingenious Britain,* Malmesbury, March 2010
[58] *The Independent,* 'Last month James Dyson said the decline of British manufacturing is a tragedy', 6 February 2002
[59] BBC News, 'Dyson to move to Far East', 5 February 2012
[60] *High value,* p. 13
[61] *Manufacturing Advantage,* p. 9
[62] BBC news, 'Pfizer to close UK research site', 1 February 2011, accessed here: http://www.bbc.co.uk/news/business-12335801
[63]
http://www.obn.org.uk/obn_/news_item.php?r=3JES5DIKAK1686, copy of the original
[64] M. Porter, www.bis.gov.uk
[65] *'Science, Innovation and the Economy'*
[66] Local Government website:
http://www.idea.gov.uk/idk/core/page.do?pageId=11239298
[67] *'Transforming the British economy'*
[68]
http://www.communities.gov.uk/newsstories/newsroom/1626460

Conclusion

[1] 'The return of the US manufacturer'

FIGURES

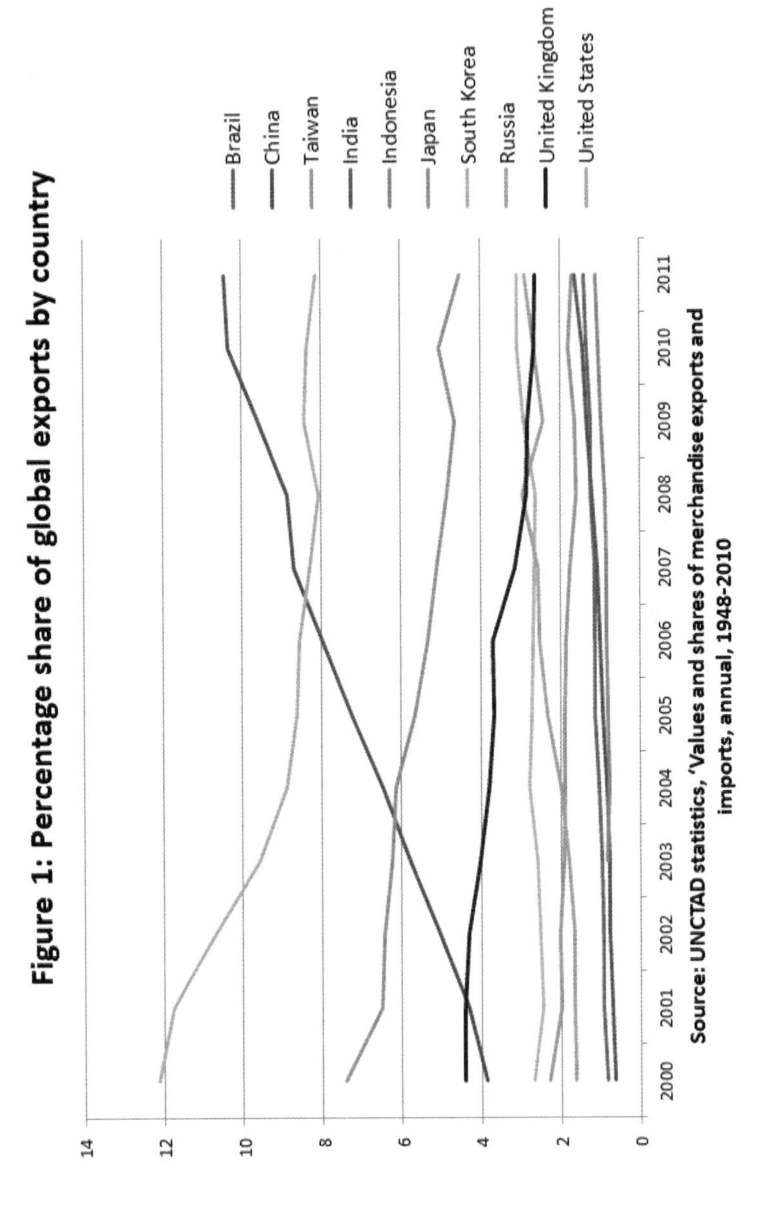
Figure 1: Percentage share of global exports by country
14
12
10
8
6
4
2
0
2000
2001
2002
2003
2004
2005
2006
2007
2008
2009
2010
2011
Brazil
China
Taiwan
India
Indonesia
Japan
South Korea
Russia
United Kingdom
United States
Source: UNCTAD statistics, 'Values and shares of merchandise exports and imports, annual, 1948-2010

Figure 2: Increase in UK manufacturing productivity
2000 = 100

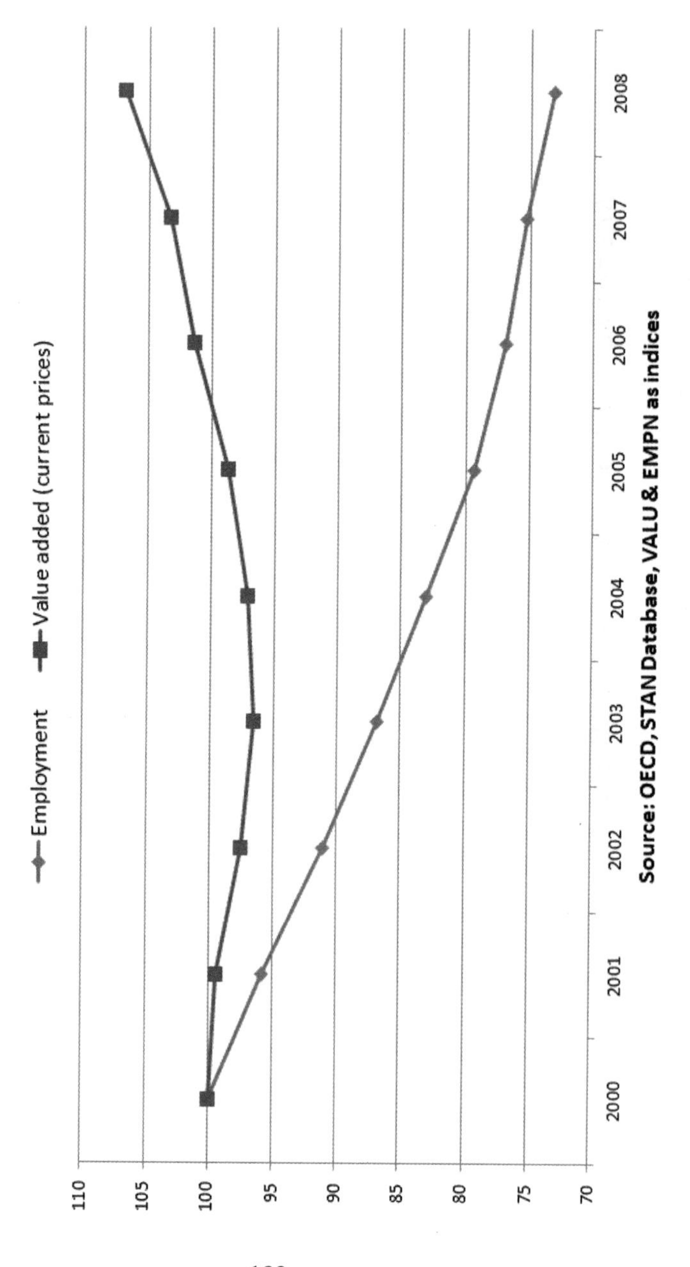

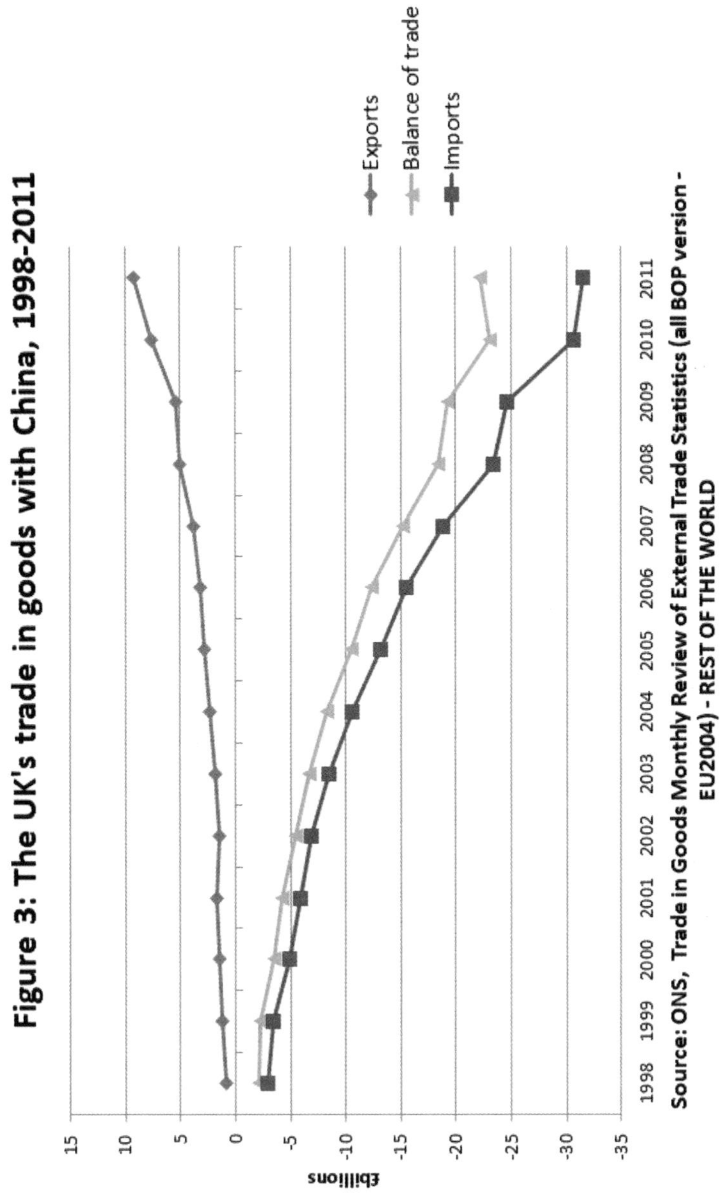

Figure 3: The UK's trade in goods with China, 1998-2011

£billions

15
10
5
0
-5
-10
-15
-20
-25
-30
-35

1998 1999 2000 2001 2002 2003 2004 2005 2006 2007 2008 2009 2010 2011

Exports
Balance of trade
Imports

Source: ONS, Trade in Goods Monthly Review of External Trade Statistics (all BOP version - EU2004) - REST OF THE WORLD

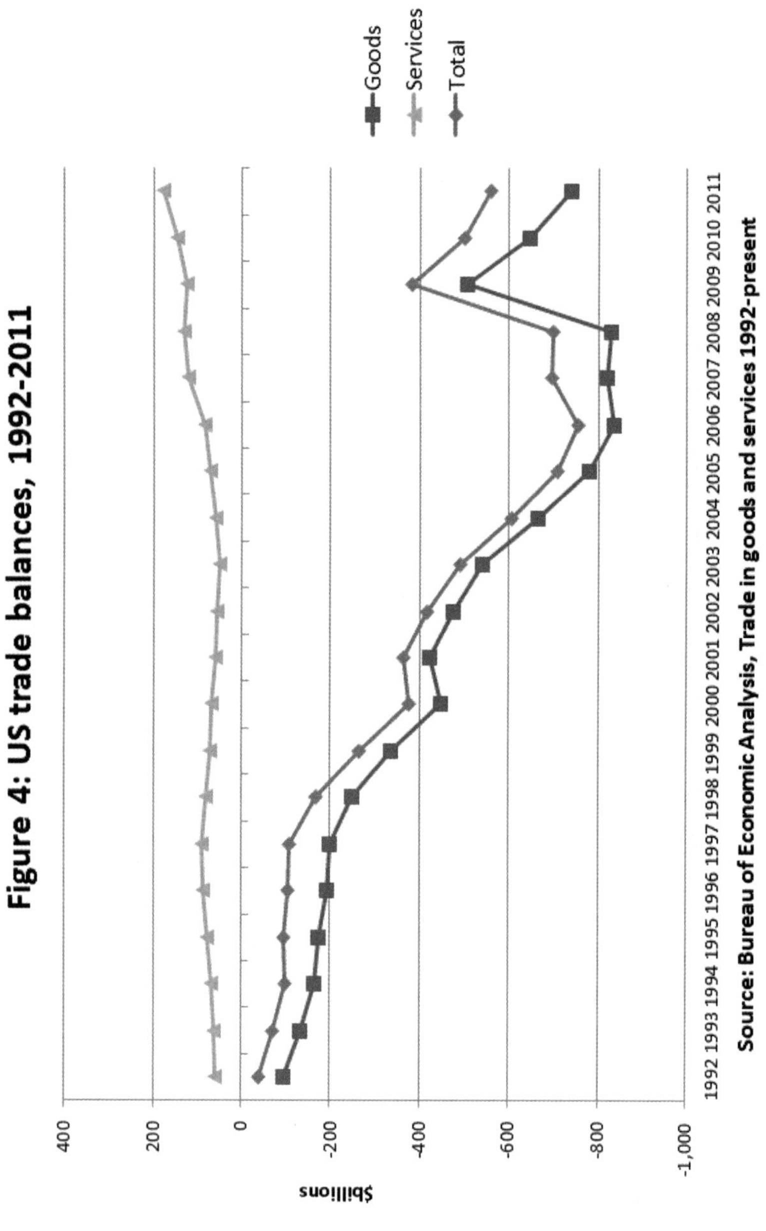

Figure 4: US trade balances, 1992-2011

$billions

400
200
0
-200
-400
-600
-800
-1,000

1992 1993 1994 1995 1996 1997 1998 1999 2000 2001 2002 2003 2004 2005 2006 2007 2008 2009 2010 2011

Goods
Services
Total

Source: Bureau of Economic Analysis, Trade in goods and services 1992-present

Figure 5: The US manufacturing workforce

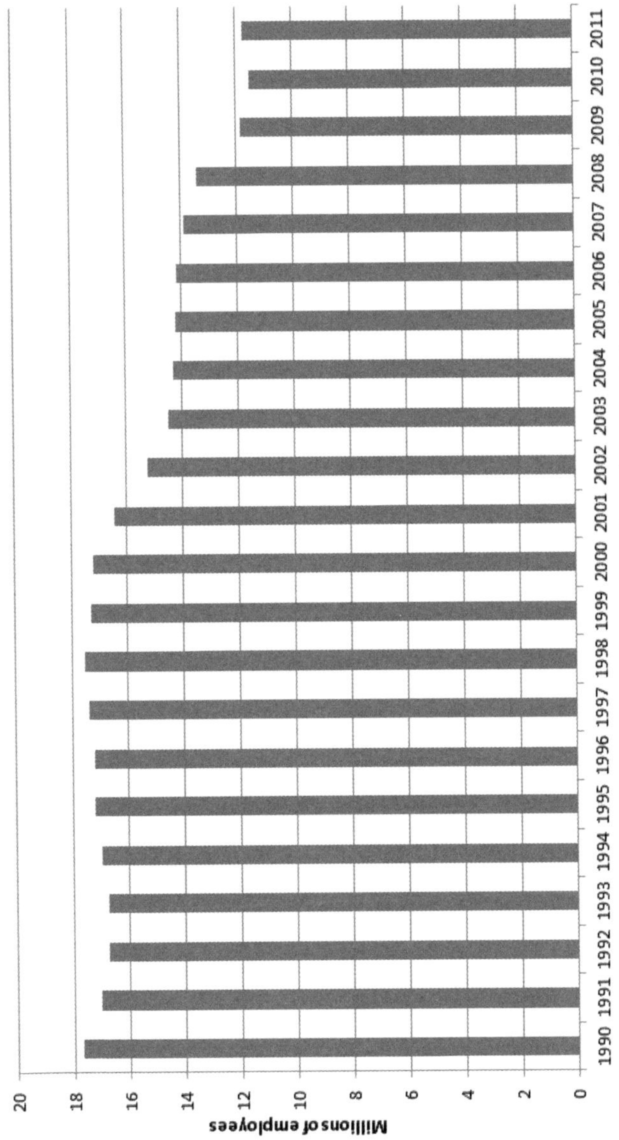

Source: Bureau of Labour Statistics, Historical employment, B-1: employees on nonfarm payrolls by major industry sector